Green Innovation and Future Technology

Other Palgrave Pivot titles

Brian M. Mazanec and Bradley A. Thayer: **Deterring Cyber Warfare: Bolstering Strategic Stability in Cyberspace**

Amy Barnes, Garrett Brown and Sophie Harman: **Global Politics of Health Reform in Africa: Performance, Participation and Policy**

Densil A. Williams: **Competing against Multinationals in Emerging Markets: Case Studies of SMEs in the Manufacturing Sector**

Nicos Trimikliniotis, Dimitris Parsanoglou and Vassilis S. Tsianos: **Mobile Commons, Migrant Digitalities and the Right to the City**

Claire Westall and Michael Gardiner: **The Public on the Public: The British Public as Trust, Reflexivity and Political Foreclosure**

Federico Caprotti: **Eco-Cities and the Transition to Low Carbon Economies**

Emil Souleimanov and Huseyn Aliyev: **The Individual Disengagement of Avengers, Nationalists, and Jihadists: Why Ex-Militants Choose to Abandon Violence in the North Caucasus**

Scott Austin: **Tao and Trinity: Notes on Self-Reference and the Unity of Opposites in Philosophy**

Shira Chess and Eric Newsom: **Folklore, Horror Stories, and the Slender Man: The Development of an Internet Mythology**

John Hudson, Nam Kyoung Jo and Antonia Keung: **Culture and the Politics of Welfare: Exploring Societal Values and Social Choices**

Paula Loscocco: **Phillis Wheatly's Miltonic Poetics**

Mark Axelrod: **Notions of the Feminine: Literary Essays from Dostoyevsky to Lacan**

John Coyne and Peter Bell: **The Role of Strategic Intelligence in Law Enforcement: Policing Transnational Organized Crime in Canada, the United Kingdom and Australia**

Niall Gildea, Helena Goodwyn, Megan Kitching and Helen Tyson (editors): **English Studies: The State of the Discipline, Past, Present and Future**

Yoel Guzansky: **The Arab Gulf States and Reform in the Middle East: Between Iran and the "Arab Spring"**

Menno Spiering: **A Cultural History of British Euroscepticism**

Matthew Hollow: **Rogue Banking: A History of Financial Fraud in Interwar Britain**

Alexandra Lewis: **Security, Clans and Tribes: Unstable Clans in Somaliland, Yemen and the Gulf of Aden**

Sandy Schumann: **How the Internet Shapes Collective Actions**

Christy M. Oslund: **Disability Services and Disability Studies in Higher Education: History, Contexts, and Social Impacts**

palgrave▸pivot

Green Innovation and Future Technology: Engaging Regional SMEs in the Green Economy

Edited by

Felicity Kelliher
Waterford Institute of Technology, Ireland

and

Leana Reinl
Waterford Institute of Technology, Ireland

Selection and Editorial matter © Felicity Kelliher and Leana Reinl 2015
Individual chapters © their contributors 2015
Foreword © Micheál Ó Cinnéide 2015
All rights reserved. No reproduction, copy or transmission of this publication may be made without written permission.

No portion of this publication may be reproduced, copied or transmitted save with written permission or in accordance with the provisions of the Copyright, Designs and Patents Act 1988, or under the terms of any licence permitting limited copying issued by the Copyright Licensing Agency, Saffron House, 6–10 Kirby Street, London EC1N 8TS.

Any person who does any unauthorized act in relation to this publication may be liable to criminal prosecution and civil claims for damages.

The authors have asserted their rights to be identified as the authors of this work in accordance with the Copyright, Designs and Patents Act 1988.

First published 2015 by
PALGRAVE MACMILLAN

Palgrave Macmillan in the UK is an imprint of Macmillan Publishers Limited, registered in England, company number 785998, of Houndmills, Basingstoke, Hampshire, RG21 6XS.

Palgrave Macmillan in the US is a division of St Martin's Press LLC, 175 Fifth Avenue, New York, NY 10010.

Palgrave Macmillan is the global academic imprint of the above companies and has companies and representatives throughout the world.

Palgrave® and Macmillan® are registered trademarks in the United States, the United Kingdom, Europe and other countries.

ISBN: 978-1-137-47981-5   EPUB
ISBN: 978-1-137-47982-2   PDF
ISBN: 978-1-137-47980-8   Hardback

A catalogue record for this book is available from the British Library.

A catalog record for this book is available from the Library of Congress.

www.palgrave.com/pivot

DOI: 10.1057/9781137479822

Contents

Foreword   viii
Micheál Ó Cinnéide

Preface   x

Notes on Contributors   xiii

Introduction   1
    Engaging regional SMEs in the green economy   3
    Supporting the enhancement of green innovativeness in regional SMEs   5
    Green Innovation and Future Technology programme overview   7

1   The Green Economy   12
    1.1   A changing economic environment   14
    1.2   Defining the green economy   15
    1.3   The policy gap   17
    1.4   The green business landscape in Ireland and Wales – a regional perspective   21

2   The Green Innovation and Future Technologies (GIFT) Concept   27
    2.1   Green innovativeness and green capability development – an SME perspective   28
    2.2   Embedding cross-border innovation and knowledge exchange in a green learning community   30

2.3		Proposing a framework of regional SME engagement for sustainable development	32
2.4		Concluding remarks	36

3 **Multilevel Engagement: Theory and Practice Integration**   41
- 3.1 Collaborative spaces to promote cross-border SME/stakeholder engagement and knowledge exchange   43
  - 3.1.1 Case A: Creating a VLE to embed a CoP ethos   46
  - 3.1.2 Case B: Training days: collaborative engagement to promote sustainable walking tourism   48
  - 3.1.3 Case C: Annual learning showcase and study tour   51
- 3.2 Concluding remarks   53

4 **Reciprocal Knowledge-Transfer Activities between SMEs and Academia**   57
- 4.1 Postgraduate – regional SME – engagement: collaborative applied research projects   60
  - 4.1.1 Business-led maxi projects   63
    - 4.1.1.1 The potential adoption of eco-trolleys by supermarket retailers in Ireland   63
    - 4.1.1.2 The effect of diet on rumen microbes and methane emissions   65
    - 4.1.1.3 Attitudes towards eco-certification in the hospitality industry in North West Wales   68
    - 4.1.1.4 Developing a green audit for Oceanics Surf School & Marine Education Centre   70
  - 4.1.2 Business-led mini projects   73
    - 4.1.2.1 Bachelor of Science (BSc) in Small Enterprise Management   73
    - 4.1.2.2 MBA student pitch to SME owners   76
  - 4.1.3 Case D: Training days: collaborative engagement to promote wood fuel quality   78
- 4.2 Developing interdisciplinary curricula in sustainable development education   80
  - 4.2.1 Case E: Greening the MBA   81
  - 4.2.2 Case F: Physical and virtual learning spaces for cross-border student knowledge exchange: green technology postgraduate study tour   84
- 4.3 Concluding remarks   86

5	Green Stakeholder Engagement: The Learning Journey	91
5.1	Reflections on cross-border interdisciplinary academic-practice collaboration	92
	5.1.1 Academic team	92
	5.1.2 Advisory board	93
	5.1.3 The support hub	97
	5.1.4 Expert/industry stakeholders	99
	5.1.5 Participating SMEs	100
	5.1.6 Participating students	104
5.2	Where next? Key insights and future directions	106

Index 114

Foreword

The green innovation and future technologies (GIFT) project was a timely initiative, aimed at up-skilling SME businesses within the INTERREG regions of Ireland and Wales to help grow a sustainable green economy. It brought together a blend of academic, policy and business perspectives from Ireland and Wales on an exploration of green growth at the local and micro business level. In documenting and assessing this learning, this publication adds to the value of the GIFT project and locates the challenges in a wider context.

The GIFT project was conceived between 2006 and 2008, during the later stages of the Celtic Tiger economic boom in Ireland, when economic growth led to a phase of unsustainable consumption. The success of the GIFT project, as charted in this book, suggests that its sustainable green economy principles were in tune with many of the emerging lessons in our societies, including

▸ the need to rebuild our economy in a way which is resource efficient, low carbon and socially inclusive;
▸ the crucial role of small and medium enterprises as a binding element in sustainable communities;
▸ the importance of valuing natural capital and ecosystems, which underpin so many activities such as coastal tourism, agriculture, forestry and the provision of energy; and
▸ the role of networking and informal social learning in adapting to change.

A pattern which emerges in the book is the value of diverse learning from multidisciplinary teams spanning business, science, engineering and sociology. The breadth of experience amongst the 23 contributors to this publication is most impressive. Institutes of higher learning can play a key role in championing regional development – while Bangor University has long been a beacon of wisdom over the Menai straits in north Wales, the emergence of Waterford Institute of Technology as a hub for praxis-based knowledge is also a catalyst for change in the southeast region of Ireland.

As Minister Richard Bruton wrote in an Irish progress report on the green economy, 'delivering the full potential of the green economy for Ireland is a long term goal'. The early chapters in this book recognise and explore the 'gap from theory to implementation' in green innovation and offer a road map for nurturing the green economy at a regional level. The later chapters give a wealth of insight and experience from case studies on the learning journey. From the Irish Environmental Protection Agency (EPA) perspective, we commend the GIFT team for their creativity, energy and skill in bringing us a few more steps along the long road towards a sustainable rural economy.

Ireland has many cultural ties with our Welsh brethren, and we have much to learn from each other. Three Welshmen helped change the course of Irish history – St Patrick, Strongbow and David Lloyd George. In modern times, we still can learn lessons from the policy focus on sustainable growth that is a hallmark of the current Welsh administration. In spreading this awareness to the regional SME business sector, the GIFT project seeks to fulfil the principles of the European Regional Development Fund (ERDF) through the INTERREG 4A Ireland Wales programme, by helping to bring our regions closer together across the Irish Sea.

Micheál Ó Cinnéide, Director, Environmental Protection Agency, Wexford, Ireland.

Preface

There is an increasingly common view that there needs to be a move towards a more environmentally sustainable path of economic growth, where a value is placed on the environment when contemplating economic activity. While many speak of this 'sustainable path', there has been little debate as to how the skills required to fulfil the goals of sustainable development can be imparted within regions and, particularly, within rural business communities. The purpose of this text is to consider green innovation and future technology skill development within regional small- to medium-sized enterprises (SMEs).

While there are many definitions of 'region', when administratively defined, a region is considered an administrative division of a country, which has incumbent policies to assist processes of economic development. For example, Tuscany is a region of Italy, the South West is a region of Wales, and the South East is a region of Ireland. Based on this definition, a regional economy is one that is nested territorially beneath the level of the country, and these subeconomies are primarily serviced by SMEs. With rare exceptions, these firms provide all commercial activity in rural regions within their respective countries, and as such dictate the level of economic activity and development within these regions.

Recent studies have shown that the framework conditions under which SMEs operate, and their entrepreneurial culture, are key factors in determining the extent of SME performance and consequently their contribution to a region's macroeconomic growth. Here, framework conditions refer to a general or systemic outlook, including growth opportunities, rates of innovation and a system's investments in innovation. The focus of this book is on green innovation and future technologies, tools which ideally contribute to 'an economy that results in improved human well-being and reduced inequalities over the long term, while not exposing future generations to significant environmental risks and ecological scarcities'.

Considering the highlighted framework conditions that promote regional innovation systems, the stimulation and development of green innovation and innovativeness represent a vital resource within the rural-regional SME, and one which needs to be honed if regions are to prosper in the green economy. The term 'innovation' is defined as 'a new idea, device or method' or, in the context of this book, 'the act or process of introducing new ideas, devices or methods'. Thus, green innovation is the introduction of green-focused ideas, devices or methods relating to regional SMEs, while green innovativeness is defined as a 'transformational innovative capability' within these firms.

A regional innovation system combines the focus of regions with a systems perspective, such that the region can evolve when systemic innovations are invoked. In context, regional economic development entails the creation of new business and/or the expansion of existing firms, while strategies for growth include improved entrepreneurial activity, enhanced entrepreneurial culture and optimised framework conditions within the region. Therefore, it is these innovation systems that provide for the sustained economic well-being of the regions they service.

In order to access green-focused ideas, devices or methods, a vital part of SME managerial work is the stimulation and development of innovation as a resource and innovativeness as an evolving capability within the firm. It is in this context that collaborative interaction is seen as a priority driver in green innovativeness, as inherent SME skill requirements will cross the boundaries of green innovation and future technologies in the natural-physical sciences and business-technology management scholarship. In this book, we contend that pursuit of such green skill enhancement should be based on cross-disciplinary collaborative action within an academic-practice partnership ethos.

In adherence to the principles of collaborative action, an interdisciplinary team of academics and project specialists crossing Schools of Science, Engineering, Business and Adult Education came together in

Wales and Ireland as part of a cross-border INTERREG-funded initiative titled 'green innovation and future technology' (GIFT).[1] This initiative sought to pursue the design, development and delivery of a cross-border collaborative green innovativeness education and partnership programme in interaction with the wider regional business community in each domain. By considering the dynamics involved in the collective up-skilling of regional actors, including SMEs, social enterprises, public sector and educational institutions, the ultimate goal was to engage an international community of 'green' practice.

Note

1 The European Regional Development Fund (ERDF) through the Ireland Wales Programme (INTERREG 4A) has provided valuable funding for the GIFT programme initiative.

Notes on Contributors

Felicity Kelliher is Senior Lecturer in Management Studies at the School of Business, Waterford Institute of Technology (WIT), Ireland. She lectures on leadership, organisational behaviour, change management and reflective/critical thinking on Doctoral, Master's and Executive programmes and has been a visiting professor/guest lecturer on Master's programmes in France, China, Canada and Ireland. As a founder and senior researcher with the RIKON research group WIT (www.rikon.ie), Felicity studies small/micro-firm learning networks and management capability development on a regional basis and has received a number of international awards for this work. A Fulbright scholar, Felicity has published in a number of international journals, including the *Journal of Tourism Management*, the *Journal of Business Ethics*, the *Journal of Small Business and Enterprise Development* and the *Journal of Entrepreneurship and Innovation*. Felicity is a consulting academic on the GIFT project who was heavily involved in the design and development of the programme from a WIT perspective.

Leana Reinl is Business Development Officer with the GIFT project at Waterford Institute of Technology and a postdoctoral researcher with the RIKON research group (www.rikon.ie). Leana was involved in the development and delivery of a national tourism learning network initiative and subsequent degree program at WIT, while completing her Master's by Research and PhD. Leana has a strong international publication record (having published in *Tourism Management*, the *Journal of Tourism Planning* and *Development*, the *International Journal of Entrepreneurship and Innovation* and the *Journal of Small Business and Enterprise Development*) with research interests in small/micro-firm learning, learning networks and evolving learning communities. Leana co-supervises an Institute of Technology Ireland-funded research project with colleagues in WIT, Bangor University, Wales, and the University of Guelph, Canada, and she will embark on a three-year Irish Research Council-funded postdoctoral fellowship to the University of Guelph in October 2014 to explore a social learning framework for rural tourism development.

Stuart Bond is the GIFT Project Manager and has worked in the Green Economy field for the past two decades. Prior to this, he was WWF-UK's Head of Research and Metrics leading on Sustainable Production and Consumption and Carbon, Ecological and Water Footprints. Equally comfortable founding local sustainability charities or engaging in global sustainability processes, Stuart has successfully originated, fundraised and managed over £3million worth of green economy projects. He is author of many influential reports including 'One Planet Wales' and 'Counting Consumption', and is a Bangor University Alumni in Forestry and Countryside Management.

Michael Breen completed his PhD in Physical Chemistry at Waterford Institute of Technology in conjunction with the Materials and Surface Science Institute at the University of Limerick. Michael is a Lecturer in the Department of Chemical and Life Sciences at WIT and programme leader for the BSc in Agricultural Science. His current research interests include the application of surfactants to the area of 'Green Chemistry' and investigating the possible use of waste streams as antioxidant sources. As well as being a member of the GIFT research team, Michael is also part of the EIRC (Eco-Innovation Research Centre) and PMBRC (Pharmaceutical and Molecular Biotechnology Research Centre) in WIT.

Keeley Clayden graduated with a First Class Bachelor of Business Honours degree in Recreation and Sport Management from WIT in 2010. With a passion for the outdoors, Keeley undertook a Master of Arts degree by research studying the personality, motivations and level of involvement of land-based recreationists in the Irish uplands, graduating from WIT in 2012. Keeley was appointed as GIFT's Sustainable Tourism Theme Officer, where her research primarily focused on developing sustainable walking tourism networks and services whilst ensuring that GIFT members developed an appreciation of the importance of protecting the natural landscape whilst developing their business in this sector. In this role Keeley liaised with stakeholders, SMEs and community/development groups, organising workshops, conferences, study tours and online discussions to support sustainable tourism in the green economy.

Evelyn Doyle is Senior Lecturer at the School of Biology & Environmental Science at the University College Dublin. Her research interests include Biodegradation and Environmental Microbiology, particularly the application of microorganisms in the degradation of xenobiotic compounds and bioremediation of contaminated sites. Recent studies focus on the novel applications of microbial enzymes, particularly those involved in xenobiotic degradation. Evelyn has published in the *Journal of Applied Microbiology*, *Environmental Biology* and the *Journal of Agricultural Science*, among others. Her role as senior academic on the GIFT programme involved the coordination of mini and maxi projects pursued by research students within GIFT programme SMEs.

Anthony Foley is Lecturer at the School of Business at Waterford Institute of Technology, and Senior Researcher in the RIKON group WIT (www.rikon.ie). His research interests are in marketing strategy, tourism and services marketing, and brand co-creation. Publications include articles in the *European Journal of Marketing* and *International Journal of Management Reviews*, and a chapter on the tourism experience brand just published in the *Routledge Handbook of Tourism Marketing*.

Heli Gittins is a Business and Skills Development Officer on the GIFT project based at Bangor University. She has developed, and teaches, environmental modules of the Environmental Management MBA and works closely with businesses developing and delivering training. She has worked on environmental knowledge-transfer projects at Bangor University for the past five years, following an MSc here in Conservation and Land Management in 2007. She enjoys the inspiration afforded by working with sustainability pioneers across the sectors – those businesses which take responsibility and risks ahead of the curve, blazing a trail for a groundswell of change in order to meet shared challenges. She is also a mindfulness teacher and has a research interest in Mindfulness-Based Approaches to Sustainable Development.

Ray Griffin is Lecturer in Management and Organisation Studies at Waterford Institute of Technology's School of Business, and is the programme director of the MSc in Business, Innovation, Technology and Entrepreneurship (mBite) in the Department of Graduate Business. Ray researches in and around the sociology of organising, with projects on multinational corporations, fun workplaces, banks as hypermodern organisations and unemployment as just another type of work, all underway. Postgraduate students on the mBite programme participated in a joint collaborative programme with Bangor University students as an aspect of the GIFT programme.

Gareth Griffiths is the Postgraduate Director of Studies at Bangor Business School, Bangor University, and is the principal investigator for the GIFT project. He has worked at a number of Business Schools and been an external examiner at seven UK Universities. He has been an assessor for Business Schools in France, Belgium, Russia, the Netherlands and Ireland. Prior to working in academia, Griffiths was a Consultancy Manager with Hewlett Packard. He has published numerous articles in a range of journals, and his research interests include green business models, business strategy, social enterprises and renewable energy.

Denis Harrington, Head of Graduate Business at Waterford Institute of Technology, holds responsibility for directing all taught graduate business programs and executive development within the school; with teaching responsibilities for Leadership and Professional Development on Doctorate and Executive MBA programmes. He has extensive research and consultancy experience, having worked in the UK, France, China and within the Russian Federation. Denis is a visiting professor at NEOMA Business School, France and at Shanghai University, China. He is a council member of the Irish Academy of Management and the MBA Association of Ireland. His research interests focus on strategic innovation and how management processes shape the emergence of new and difficult-to-replicate strategic advantages and managerial capabilities for innovation within a tourism micro-firm context. He serves on the editorial boards of the *Service Industries Journal*, UK, the Irish *Journal of Management* and the *Journal of Applied Research in Higher Education*, USA.

Audrey Hearne is Lecturer in the Department of Chemical and Life Sciences at Waterford Institute of Technology. She is also course leader for the BSc in Applied Biology at WIT. Upon obtaining a BSc (Hons) in Microbiology at NUI, Galway she completed her PhD in Molecular Biology at WIT. Her PhD studies involved the cloning and expression of the cyanide hydratase gene into yeast with a view to assessing its potential for the bioremediation of cyanide. Audrey has been involved with a number of research groups at

WIT since completing her PhD, including the Macular Pigment Research Group (MPRG) and Biomedical Research Cluster (BRC). Her main areas of interest are environmental microbiology and molecular biology in terms of bioremediation. Currently she is a member of the collaborative partnership and research project Green Innovation and Future Technologies.

David Joyner is Executive Director of the Confucius Institute at Bangor University and is Visiting Professor in the School of Humanities, China University of Political Science and Law (CUPL), Beijing. He holds a PhD in Chemistry and has experience in academic research in Chemistry and Physics at the Universities of Wales (Swansea), Pittsburgh (USA) and Cambridge. He has worked in large and small companies, including industrial research in Unilever plc. David also established a knowledge intensive company in scientific imaging software and systems. He has 20 years of experience, having worked at the university/industry interface on knowledge transfer and business development. Extensive inter-disciplinary international collaborations have been based in fields such as creative industries, arts enterprises, green innovation, early stage knowledge intensive businesses, and social enterprises. He has a research programme in 'Creativity Transfer' between CUPL and Bangor University's School of Music where he is an Honorary Senior Research Fellow.

Maeve Kennealy is a postgraduate researcher who has worked on the GIFT project as a waste management theme officer. Maeve graduated from Waterford Institute of Technology with a BSc (Hons) in Applied Biology with Quality Management. She is currently pursuing an MSc by Research in the Eco Innovation Research Centre at WIT. Her research focuses on the development of a novel biofuel using solid pig manure. In her time with GIFT, Maeve conducted analysis in the area of the detection of polyphenols and the determination of acetyl content in wood using a number of analytical methods. Maeve has presented at, and published in the, proceedings of both national and international conferences. Her research interests include biomass and bioenergy processes and utilisation, and the environment.

Tom Kent is the Course Leader of the BSc in Forestry at Waterford Institute of Technology and the project co-ordinator in the Department of Agriculture Food and Marine, COFORD-funded Forest Energy Research Programme. The Forest Energy Research Programme 2010–2014 aims to investigate cost-effective wood fuel supply chains from Irish forests to meet the growing demand for indigenous, renewable, carbon-neutral energy. Tom is a member of the oversight committee of the Wood Fuel Quality Assurance Scheme, and WIT is contracted to carry out independent verification testing for the scheme. Tom was involved in the coordination of GIFT's 'Wood Fuel Quality Management' event where he, alongside a number of research colleagues at WIT, presented insights on managing wood fuel quality along the supply chain.

Morag McDonald holds a personal chair in ecology and catchment management at Bangor University. She is Head of the School of Environment, Natural Resources and Geography, which has 400 undergraduate students, 162 Master's students, 60 PhD students and 28 members of faculty. She has been an active researcher in international environmental issues for over 20 years, with broad research expertise in soil conservation and fertility; impacts of anthropogenic and natural disturbance on forest ecosystems; tropical forest regeneration; agroforestry systems; water regulating ecosystem services and forest restoration through fallow management. She has field experience in 14 countries. She is a founder and consortium member of the joint European MSc programme in Sustainable Tropical Forestry (SUTROFOR), and the global PhD programme in Forest and Nature for Society (FONASO). She is an associate of the Institute of Environmental Management and Assessment, and a fellow of the Royal Geographical Society.

Jane Russell-O'Connor lectures at Waterford Institute of Technology in both the Departments of Education and Architecture and has been teaching in a range of fields for 19 years. Her research interests are principally in landscape ecology and landscape characterisation, as well as more general environmental issues such as environmental education. Her prior research has been both national and international. Prior to teaching, she ran a nature reserve, an environmental education centre and a country park. Her current research focuses on the ecological and historical features of the landscapes of Irish landed estates.

Eleanor Owens is Lecturer in the Chemical and Life Sciences Department of WIT, and a project co-ordinator for GIFT. Before entering academia she worked as a senior research officer at the Paint Research Association and as product manager, De La Rue, both in the UK. Since joining WIT in 2002, Eleanor has lectured in Physical Chemistry, Applied Chemistry, Quality Management and Science Education at both undergraduate and postgraduate level. She is an active researcher with research interests in wood energy, novel biomass products and antioxidants. She is also a member of the PMBRC (www.pmbrc.org), and is working on a number of industry-sponsored projects investigating drug delivery platforms.

Sean Storey holds a BSc in Microbiology and a PhD in Microbial Ecology, both from University College Dublin. On completion of his PhD, Sean was appointed GIFT research assistant at UCD, where he is part-responsible for managing the University's input to the GIFT project. His research interests lie in the biodegradation of xenobiotic compounds and in the use of biological systems to remediate brownfield sites. He is a member of both the Society for General Microbiology and the International Society for Microbial Ecology, and his work has been published in peer-reviewed scientific journals.

John Wall has over 25 years of work experience in a number of settings. He has worked in industry as a civil engineer and as a lecturer, researcher and course administrator in Waterford Institute of Technology. As Head of School of Lifelong Learning and Education, John heads up a team that supports access to part-time undergraduate courses as well as supports education-focused courses that address Teaching Council requirements for tutors in the further education sector. As a lecturer in WIT, John led the development of a new blended learning course focused on the lifelong learning needs of construction professional, one of the first blended courses in WIT. As a research coordinator, John has been project manager of an EU Minerva-funded research project, focused on the continuing professional development needs of construction professionals in Europe.

James Walmsley is Lecturer in Forestry in the School of Environment, Natural Resources and Geography. He is director for two highly successful distance-learning courses: MSc Forestry and MSc Tropical Forestry. These courses combine Bangor University's impressive reputation and expertise in teaching forestry with the latest green technologies to deliver high-quality teaching and learning to students with limited access to postgraduate level study. James has published in a number of leading forestry journals, and his research interests include biomass and wood energy, forest inventory and measurement, social aspects of forestry, and household energy consumption and behaviour. He has received a number of awards, including the 2010 Silvicultural Prize of the Institute of Chartered Foresters in 2010. In 2014 he was awarded a teaching fellowship in recognition of his commitment to teaching and pastoral care at Bangor University.

Margaret Walsh is a postgraduate researcher working on the GIFT project. She graduated from WIT with a Bachelor of Arts Honours degree in

xx   Notes on Contributors

Languages and Marketing (French and Italian) in 2005, and a Master's in Business Studies, specialising in marketing, in 2006. She is currently pursuing a PhD with the RIKON group and under the supervision of Dr Patrick Lynch and Prof. Denis Harrington. Her PhD research focuses on firm-level innovativeness in an Irish SME tourism context. Margaret has extensively published in the proceedings of national and international peer-reviewed conferences and in academic journals, including the *Journal of Business Research*, the *International Journal of Management Review* and the *Irish Journal of Management*. Margaret is a graduate member of the Marketing Institute of Ireland. Her research interests lie in the field of organisational innovation, strategic management and tourism marketing.

Einir Young is Director of Sustainability at Bangor University, and with her group SBBS (Synnwyr Busnes – Busines Sense www.sbbs.org.uk) leads on sustainability issues across the institution. She is the Director of WISE Network at Bangor (www.wisenetwork.org), a cross-disciplinary ERDF-funded collaboration between Bangor, Aberystwyth and Swansea Universities, working with businesses to take advantage of the green economy by developing sustainable products, processes and services. The SBBS *Healthcheck* is recognised as the most innovative and successful way of introducing cross-cutting themes of environment management and equal opportunities to business. Einir chairs the Gwynedd Environment Partnership, and is the academic lead on a project to successfully establish the first Ecomuseum in Wales (www.ecoamgueddfa.org/). A leading expert in the field of sustainability, her main interest is in institutional sustainability and integrated sustainable business development and reporting.

Introduction

F. Kelliher and L. Reinl

Abstract: *Kelliher and Reinl discuss the links between economy and environment, before expanding on the concept of sustainable business development in the emerging green economy. The authors go on to focus on regional small to medium-sized enterprises (SMEs), and the approaches that can be taken by business, state, semi-state, education and community stakeholders to engage SMEs in the green economy. The chapter debates the value of green innovativeness, before offering an overview of an incumbent regional stakeholder engagement development programme. This green innovation and future technology (GIFT) programme pursues the potential up-skilling of regional SMEs' green innovativeness via multidisciplinary cross-regional/ national support structures and in doing so, assists in the development of the green economy.*

Kelliher, Felicity and Leana Reinl. *Green Innovation and Future Technology: Engaging Regional SMEs in the Green Economy.* Basingstoke: Palgrave Macmillan, 2015. DOI: 10.1057/9781137479822.0005.

Since the 'blueprint for a green economy' was first published in 1989, the link between economy and environment has captured global attention (Cai, Wang, Chen and Wang, 2011; Da Graça Carvalho, Bonifacio and Dechamps, 2011). As highlighted in the preface, the common view is that there needs to be a move towards a more environmentally sustainable path of economic growth (Da Graça Carvalho et al., 2011; Kucera, 2009), in which a value is placed on the environment when contemplating economic activity (Pearce, Markandya and Barbier, 1989). It is the authors' contention that this 'sustainable path' requires the development of green innovation skills to fulfil the goals of sustainable development within regions (Renner, Sweeney and Kubit, 2008) and particularly, within rural business communities.

Increasingly, political support is founded on an input-output model of engagement, as reflected in recent policy documents and statements released in Ireland (2011), the UK[1] (2011), Wales (2009) and Europe (European Environment Agency, 2013), each of which posits a government objective of creating a resource-efficient and smart, green economy. This ethos is reinforced by the United Nations Environment Programme (UNEP) (2011, p. 9), which defines the green economy as

> *A system of economic activities related to the production, distribution and consumption of goods and services which will result in improved human well-being over the long term, while not exposing future generations to significant environmental risks or ecological scarcities...a green economy can be thought of as one which is low carbon, resource efficient and socially inclusive.*

This definition is supported by the European Union (EU) sustainable development strategy framework (2009), and the European environment policy targets and objectives (European Environment Agency, 2013) as it provides for a long-term vision of sustainability in which economic growth, social cohesion and environmental protection go hand in hand and are mutually supporting. Thus, governments, policy officials and industry leaders are all calling for businesses to place a greater emphasis on producing green innovations through more sustainable means in order to contribute to the goal of a greener economy (OECD, 2009).

When considering European businesses, SMEs are defined as having no more than 250 employees and a turnover of less than €50 million per annum (European Commission, 2011). The vast majority of European economies are made up of regions dominated by SMEs (Cox et al., 2013): collectively, 20 million SMEs represent 99.8 per cent of businesses in the European Community (EC), employing 86.8 million people and contributing €3.4 trillion to the economy on an annual basis (Cox et al., 2013). Therefore, the focus of this text is on regional innovation systems (Cooke and Leydesdorff, 2006), and specifically, the pursuit of regional SME green innovativeness skill enhancement and ultimately, a greener economy (OECD, 2009).

While there are many definitions of 'region', when administratively defined, a region is considered an 'administrative division of a country', which has incumbent 'policies to assist processes of economic development' (Cooke and Leydesdorff, 2006, p. 6). Based on this definition, a regional economy is one that is 'nested territorially beneath the level of the country' (p. 6). As highlighted in the preface, SMEs provide virtually all rural-regional commercial activity in the EC (Cox et al., 2013), and thus SME engagement in the green economy is paramount in order to achieve EC goals in this arena.

Engaging regional SMEs in the green economy

Recent studies have shown that the 'framework conditions SMEs operate under and their entrepreneurial culture are key factors in determining the extent of SME performance and consequently their contribution to macroeconomic growth' (Franks et al., 2010; Regtering et al., 2013, as cited in Cox et al., 2013, p. 11). Here, framework conditions refer to a general or systemic outlook, including growth opportunities, rates of innovation and a system's investments in innovation. The term 'innovation' is defined as 'a new idea, device or method', or in the context of this book, 'the act or process of introducing new ideas, devices or methods' (*Merriam-Webster Dictionary*). Thus, green innovation is the introduction of green-focused ideas, devices or methods relating to regional SMEs, while green innovativeness is defined as a 'transformational innovative capability' (Dutta, Narasimhan and Rajiv, 2005; Lado and Wilson, 1994; Wang and Ahmed, 2007) within these firms.

At a macro-level, a 'regional innovation system combines the focus of regions with a systems perspective' as the 'trajectory of a region can be the subject of evolution when systemic innovations are involved' (Cooke and Leydesdorff, 2006, p. 5). In context, regional economic development entails the creation of new business and/or the expansion of existing firms, while strategies for growth include improved entrepreneurial activity, enhanced entrepreneurial culture and optimised framework conditions within the region. Therefore, it is these innovation systems that provide for the sustained economic well-being of the regions they service, and in the context of this text, through regional SME engagement in the green economy.

Notwithstanding the goals of a greener Europe (OECD, 2009), there have been challenges in encouraging green innovation activities within the regions, as regional SMEs continue to pursue 'economic activities related to the production, distribution and consumption of goods and services' (UNEP, 2011, p. 9) that are often at odds with the green ethos. This is the case despite findings that the stimulation and development of green innovation and innovativeness are a vital resource within rural-regional SMEs (Santoro and Chakrabarti, 2002; Walsh et al., 2012), and need to be honed if regions are to prosper in the green economy.

The anomaly between what European policy seeks and what regional suppliers offer may be partly explained by how regional SMEs operate. SMEs face very different resource challenges to their larger counterparts (Kaufmann and Tödtling, 2002; Kelliher and Reinl, 2009), particularly in relation to the development of a green innovativeness skill set (Santoro and Chakrabarti, 2002; Walsh et al., 2012). As the majority of regional firms are on the small end of the spectrum (fewer than 50 employees), owners are mainly concerned with the day-to-day running of the business (Kelliher et al., 2014; Storey and Cressy, 1996). Regional firms also tend to build competitive advantage based on localised knowledge (Wickham, 2001), rather than focusing on an evolving national or international perspective. Therefore, if the regional market is not *currently* demanding a green deliverable, or willing to pay a premium for this offering, the SME is unlikely to pursue 'green-focused ideas, devices or methods', thereby creating a scenario in which green innovativeness lies largely untapped within these firms.

This *current* activity emphasis coupled with resource constraints often manifests itself in SME owners who are focused on immediately applicable performance (Freel, 1999) to the detriment of long-term skill enhancement. This reality is further compounded by the SME owners' role of primary decision-maker, as they often depend on their own intuition over formal decision models when considering the strategic direction of the firm (Rice and Hamilton, 1979). The owner may therefore rely on their own opinion rather than external expertise or market research to assess the need for green innovativeness within their business.

These indicators suggest specific barriers to green innovativeness skill enhancement within regional SMEs, which will need to be addressed in appropriate interventions in pursuit of a green economy.

The challenge for SMEs is that they need to invest in skill enhancement to future-proof their business, while simultaneously generating sufficient revenue to provide a reasonable economic return for their efforts. As skill enhancement takes time, investing in skill development may not appeal to an owner who is focused on immediately applicable performance (Freel, 1999). Unfortunately, the *current* focus described above can leave the SME exposed to both policy and market shifts if it focuses only on the 'now'. Thus, despite market closeness, the SME may not be aware of, or take due care during a policy shift, as is happening currently, based on the proposed introduction of penalties for non-conformance of green criteria at European level (European Environment Agency, 2013). As a result, the regional SME may take some time to respond to these dynamics, resulting in penalties imposed for breach of green policy and/ or loss of competitive advantage as consumers become more aware of what they should expect from a regional supplier. Therefore, the SME dilemma is how to remain environmentally responsible and sustain economic success, while reducing the burden of commercial activities on the current and future environment. This challenge places a significant resource burden on regional SMEs and therefore requires specific support in honing a green innovativeness skill set for this cohort.

Supporting the enhancement of green innovativeness in regional SMEs

In order to access green-focused ideas, devices or methods, a vital part of SME managerial work consists of the stimulation and development of innovation as a resource and innovativeness as a capability within the firm. It is in this context that collaborative interaction is seen as a priority driver in green innovativeness (Brooks and Ryan, 2008; Nastase, Popescu and Boghean, 2009), as inherent SME skill requirements cross the boundaries of green innovation and future technologies in the natural-physical sciences and business-technology management scholarship.

For stakeholders contemplating a regional SME green innovativeness programme, comprehensive stakeholder collaborative engagement is a vital component when pursing transformational innovative capability (Lado and Wilson, 1994; Dutta et al., 2005; Wang and Ahmed, 2007) in these firms. Specifically, government agencies, higher education institutes (HEIs), indigenous businesses, economic support groups and rural development groups are each considered pivotal to successful and sustainable regional development (Döring and Schnellenbach, 2006; Drda-Kühn and Wiegand, 2010; Kelliher et al., 2014), and should therefore be involved in a capability development programme of this nature. This is particularly important in more rural domains, where SMEs are smaller in nature and therefore operate under the aforementioned resource restrictions. Thus, this book documents the development programme journey from a cross-border, multidisciplinary perspective, and in doing so, seeks to

- identify the green skill needs of regional SMEs;
- examine how best to impart green innovativeness skills to regional SMEs;
- design an integrated approach to regional SME green skill enhancement;
- present a cross-border multiperspective lens through illustrated case studies;
- explore the benefits of a stakeholder approach to green innovation and future technology education;
- proffer a stakeholder approach within a rural context to facilitate greater knowledge exchange amid cross-border communities of practice;[2] and
- engage a cross-border, multidisciplinary philosophy in seeking to create a resource-efficient and smart, green economy.

This book provides an overview of the emerging global green economy before presenting a detailed chronology of cross-border multidisciplinary engagement of regional SMEs in green innovation and future technologies in order to encourage SME interaction with the green economy. It offers insight into an integrated approach to SME green innovativeness skill enhancement, and includes illustrative cases to highlight the benefits of a stakeholder approach to green innovation and future technology education. The presented project is distinctive in many ways, but particularly in its cross-border, multidisciplinary approach to the green innovativeness skill enhancement of SMEs, graduates, academics and other regional stakeholders. The illustrated cases map the cross-border regional stakeholder engagement approach and highlight the SMEs' green trajectory.

The authors reveal fundamental criteria in the building of an effective approach to regional SME green innovativeness skill enhancement by documenting the progressive application of a cross-border, multidisciplinary philosophy in seeking to create a resource-efficient and smart, green economy. In time, these activities should have the capacity to support the development of a green innovativeness skill set and foster reciprocal innovation and knowledge transfer via a cross-border community of practice (Lave and Wenger, 1991; Wenger, 1998). The book culminates in a proposed framework of green stakeholder engagement, of value to policymakers, academics and government support agents interested in engaging SMEs in regional innovation systems and cross-border green innovation fora.

Green Innovation and Future Technology programme overview

In adherence to the principles of collaborative action, an interdisciplinary team of academics and project specialists crossing the Schools of Science, Engineering, Business and Adult Education in two countries, came together as part of a cross-border INTERREG-funded initiative titled 'green innovation and future technology' (GIFT). The goal was to engage with and up-skill regional SMEs' green innovativeness and in doing so, assist in the development of the green economy in each region: North West and South West Wales and the South East and East of Ireland. As a cross-border exercise, the programme sought to contribute to European-level policy (Cox et al., 2013; European Environment Agency, 2013) in relation to regional SME engagement with the emerging green economy.

The success of the GIFT programme hinged on integrated rural-regional stakeholder engagement (Döring and Schnellenbach, 2006; Drda-Kühn and Wiegand, 2010); therefore, the programme directors sought the active participation of regional, national and international stakeholders in this cross-border multidisciplinary project. At design stage, this programme had a number of challenges associated with it. There is little research on cross-border innovation fora which deal with regional/rural innovation systems and spaces across Europe (Trippl,

2010), so relevant reference points were few and far between. In addition, cross-border innovation research is primarily focused on a multinational corporation perspective (Zander and Sölvell, 2000), which does not serve regional stakeholder communities, particularly those dominated by SMEs. Therefore, the GIFT directors and team members knew that expertise was required on regional economics, rural SMEs, green innovation, green technology and regional support structures in order to ensure the programme provided the sought-for SME engagement with the emerging green economy.

Once the academic team had been gathered, a first step in this journey was to contemplate the component elements of the 'green economy' and how the green ethos and incumbent criteria interacted within and among European regions. The insights from this first phase of the GIFT programme are documented in Chapter 1.

Notes

1 The UK Government announced the establishment of a Green Investment Bank, designed to accelerate private sector investment in the UK's green economy (http://www.bis.gov.uk/news/topstories/2011/Dec/green-investment-bank-takes-a-step-closer/) [Accessed 13 March 2014].
2 As a most simplistic descriptor, communities of practice comprise a group of people who share a concern for the pursuit of knowledge or activity of some kind, and following that goal, they interact with one another regularly (Lave and Wenger, 1991).

References

Cai, W., Wang, C., Chen, J. and Wang, S. 2011. Green economy and green jobs: myth or reality? The case of China's power generation sector, *Energy*, 36, 5994–6003.

Cooke, P. and Leydesdorff, L. 2006. Regional development in the knowledge-based economy: the construction of advantage, *Journal of Technology Transfer*, 31, 5–15.

Cox, D., Gagliardi, D., Monfardini, E., Cuvelier, S., Vidal, D., Laibarra, B., Probst L., Schiersch, A. and Mattes, A. (Editors) 2013. A recovery on the horizon? Annual Report on European SMEs 2012/2013. European Commission, October.

Da Graça Carvalho, M., Bonifacio, M. and Dechamps, P. 2011. Building a low carbon society, *Energy*, 36, 1842–1847.

Döring, T. and Schnellenbach, J. 2006. What do we know about geographical knowledge spillovers and regional growth? A survey of the literature, *Regional Studies*, 40(3), 375–395.

Drda-Kühn, K. and Wiegand, D. 2010. From culture to cultural economic power: rural regional development in small German communities, *Creative Industries Journal*, 3(1), 89–96.

Dutta S, Narasimhan, O. and Rajiv S. 2005. Conceptualizing and measuring capabilities: methodology and empirical application, *Strategic Management Journal*, 26(3), 277–285.

European Commission, 2011 *Observatory of European SMEs*. [Internet] Available at: http://ec.europa.eu/enterprise/policies/sme/facts-figures-analysis/index_en.html [Accessed 12 March 2014].

European Environment Agency. 2013. *Towards a green economy in Europe: EU environmental policy targets and objectives 2010–2050*. Luxembourg: Publications Office of the European Union, EEA Report No. 8/2013 [Accessed 12 March 2014].

Freel, M.S. 1999. Where are the skills gaps in innovative small firms? *International Journal of Entrepreneurial Behaviour and Research*, 5(3), 144–154.

Ireland. 2011. *A Framework for Sustainable Development for Ireland*. Dublin: Department of the Environment, Community and Local Development, pp. 1–91. [Internet] Available at: http://www.environ.ie/en/Publications/Environment/Miscellaneous/FileDownLoad,29081,en.pdf [Accessed 12 March 2014].

Kaufmann, A. and Tödtling, F. 2002. How effective is innovation support for SMEs? An analysis of the region of upper Austria, *Technovation*, 22(3), 147–159.

Kelliher, F. and Reinl, L. 2009. A resource-based view of micro-firm management practice, *Journal of Small Business and Enterprise Development*, 16(3), 521–532.

Kelliher, F., Aylward, E., Lynch, P., 2014, Exploring rural enterprise: the impact of regional stakeholder engagement on collaborative rural networks. C. Henry & G. McElwee (eds) *Exploring Rural Enterprise: New Perspectives on research, policy and practice*, London: Routledge.

Kucera, D. 2009. Green economy and green jobs: myth or reality? *Sustainable Development: A Challenge for European Research Conference Proceedings*, 26–28 May, Brussels.

Lado, A.A. and Wilson, M.C. 1994. Human resource systems and sustained competitive advantage: a competency-based perspective, *Academy of Management Review*, 19(4), 699–727.

Lave, J. and Wenger, E. 1991. *Situated Learning: Legitimate Peripheral Participation*. Cambridge: University of Cambridge Press.

OECD. 2009. Sustainable Manufacturing and Eco-Innovation: Towards a Green Economy, *Policy Brief*, June.

Pearce, D., Markandya, A. and Barbier, E. 1989. *Blueprint for A Green Economy*. London: Earthscan Publications.

Renner, M., Sweeney, S. and Kubit, J. 2008. Green Jobs: Towards Decent Work in a Sustainable, Low Carbon World, *Digest of Green Reports and Studies*, United Nations Environment Programme.

Rice, G.H. and Hamilton, R.E. 1979. Decision theory and the small businessman, *American Journal of Small Business*, 6(4), 1–9.

Santoro, M.D. and Chakrabarti, A.K. 2002. Firm size and technology centrality in industry-university interactions, *Research Policy*, 31(7), 1163–1180.

Storey, D. and Cressy, R. 1996. Small business risk – a firm bank perspective, working paper: the Centre for Small & Medium-sized Enterprises (CSME), Vol. 39, Warwick Business School, Coventry, pp. 1–16.

Trippl, M. 2010. Developing cross-border regional innovation systems: key factors and challenges, *Journal of Economic and Social Geography*, 101(2), 150–160.

UNEP. 2011. *Towards a Green Economy: Pathways to Sustainable Development and Poverty Eradication, A Synthesis for Policy Makers* (pp. 1–52), [Internet] Available at: http://www.unep.org/greeneconomy/Portals/88/documents/ger/GER_synthesis_en.pdf [Accessed 12 March 2014].

Wales. 2009. *Capturing the Potential: A Green Jobs Strategy for Wales*, Wales: Crown Publishers.

Walsh, M., Kelliher, F., Harrington, D. and Lynch, P. 2012. Moving Towards a Green Economy: Capitalising on Organisational Innovation Capability to Leverage the Reservoir of Knowledge in Learning Organisations – An Irish Perspective, IFASM Conference Proceedings, University of Limerick, 25–27 June.

Wang, C.L. and Ahmed, P.K. 2007. Dynamic capabilities: A review and research agenda, *International Journal of Management Reviews*, 9(1), 31–51.

Wenger, E. 1998. *Communities of Practice: Learning, Meaning and Identity*. New York: Cambridge University Press.

Wickham, P. 2001. *Strategic Entrepreneurship*. London: Pitman.

Zander, I. and Sölvell, O. 2000. Cross-border innovation in the multinational corporation: a research agenda, *Journal of International Studies of Management and Organization*, 30(2), 44–67.

1
The Green Economy

S. Bond, H. Gittins, G. Griffiths, D. Harrington, D. Joyner, M. McDonald, E. Owens and M. Walsh

Abstract: *Chapter 1 sets out to define the green economy within a 'post-carbon society' and contemplates the emerging policy-practice gap from a regional perspective. The authors go on to discuss the challenges and opportunities exposed by this changing economic landscape, and consider green innovativeness as a catalyst for SME-driven regional sustainability. The chapter concludes with a discussion around the green economy transition process in regional SMEs and incumbent capabilities; a long-term and forward-oriented business focus, outward-looking sustainability planning ethos, and concentration on innovation capability enhancement. Combined, these capabilities promote socially inclusive and collective commercial activities to stimulate green growth, while protecting the environment.*

Kelliher, Felicity and Leana Reinl. *Green Innovation and Future Technology: Engaging Regional SMEs in the Green Economy.* Basingstoke: Palgrave Macmillan, 2015. DOI: 10.1057/9781137479822.0006.

As highlighted in the introduction, it is generally accepted that there needs to be a move towards a more environmentally sustainable path of economic growth, and that a green economy affords the potential for mutual delivery of economic and environmental goals. As articulated by the European Environment Agency (2013, p. 5),

> the prevailing model of economic growth – founded on ever-increasing consumption of resources and emission of pollutants – simply cannot be sustained in a world of finite resources and ecosystem capacity.

The concept of the green economy has received significant international attention over recent years, and is considered an important tool in addressing sustainable growth (United Nations Department of Economic and Social Affairs, Division for Sustainable Development, 2012). The economic value of the green economy on a global scale was estimated to be $5 trillion in 2010, employing in excess of 30 million people worldwide (Ernst and Young, 2012, citing Innovas Solutions and Kmatrix), while the green sector is projected to increase to the order of $6 trillion by 2015 (Forfás, 2011, citing Innovas Solutions) with an average growth rate of 3.7 per cent per annum for the foreseeable future.

However, the green economy should not solely focus on economics, as this narrow perspective would merely replace one 'prevailing model of economic growth' with another. When contemplating a move towards a green economy, mutual delivery of economic and environmental goals is paramount. As articulated in the fifth IPCC Assessment report (2014), compelling evidence of man-made climate change, degradation of resources, negative impact of increasing greenhouse gases and pollution, and land and water degradation points to the need for significant shifts in policy and practice. Notably, the Organization for Economic Cooperation and Development's (OECD's) Environmental Outlook to 2050 (OECD, 2012, foreword, p. 3) 'baseline' scenario predicts that

> unless the global energy mix changes, fossil fuels will supply about 85 per cent of energy demand in 2050, implying a 50 per cent increase in greenhouse gas (GHG) emissions and worsening air pollution. The impact on the quality of life of our citizens would be disastrous. The number of premature deaths from exposure to particulate pollutants could double from current levels to 3.6 million every year. Global water demand is projected to increase by 55 per cent to 2050. Competition for water would intensify, resulting in up to 3.2 billion people living in severely water-stressed river basins. By 2050, global terrestrial biodiversity is projected to decline by a further 10 per cent.

These projections reinforce the need for world economies and international trade to be more sustainable and responsible, while a '... growing awareness of humanity's impact on the environment [has] pushed the "green economy" concept into mainstream policy debate in recent years' (European Environment Agency, 2013, p. 5).

A transition to a true green economy focused on mutual delivery of economic and environmental goals can help mitigate some of the issues highlighted in the above reports and help economies to move towards the exchange of environmentally friendly goods and services. This holistic approach should significantly reduce 'environmental risks and ecological scarcities' while simultaneously improving 'human well-being and social equity', thereby assisting with 'low-carbon resource efficiency' as well as generating economic opportunities that are 'socially inclusive' (UNEP, 2011, p. 16).

1.1 A changing economic environment

Contemporary debate considers a 'post-carbon society' (UNEP, 2014) and focuses on renewable energy, energy storage and smart grids as a basis for the green economy. European Union (EU) policymakers assert that eco-innovation is critical to supporting our society in the future and minimising the gap between environmental devastation and a comfortable standard and means of living. Policymakers also advocate 'green technology' perspectives aimed at acquiring and developing advanced technologies in pursuit of the economic-environment balance at the heart of the green economy. The development of that technology involves eco-innovation in products, processes and, increasingly, services, much of which is provided by small to medium-sized enterprises (SMEs) (European Commission, 2010). In context, the OECD (2011a) highlights that 'green' and 'growth' can go together, asserting that governments should put in place policies that tap into innovation, investment and entrepreneurship, thereby driving the shift towards greener economies (Fostering Innovation for Green Growth, foreword, p. 3).

Governments, policy officials and industry leaders are all calling for businesses to place a greater emphasis on producing green innovations through more sustainable means in order to contribute to the goal of a greener economy (OECD, 2011a, 2010). Political support is founded on an input-output model of engagement, as reflected in recent policy documents and statements released in Ireland (2011), Wales (2009) and Europe (European Environmental Agency, 2013), each of which posits a government objective of creating a resource-efficient and smart green economy.

A move towards a green economy has the potential to create enhanced trade opportunities for regional SMEs by opening new domestic and export markets for environmental goods and services, by increasing trade in products certified for sustainability and promoting certification-related services, and by greening supply chains. In addition, the adoption of more resource- and energy-efficient production methods has an important role to play in securing regional SME access to existing and new markets. As highlighted previously, this is already a five trillion-dollar sector (2010) and one which is set to grow to the order of six trillion dollars by 2015 (Forfás, 2011, citing Innovas Solutions).

However, there are challenges within this green economy framework. From an economic perspective, improved resource efficiency should produce increased returns, although these could remain detrimental to the environment. With reference to the ecosystem, environmental resilience cannot be at the cost of economics, considering the proposed input-output model and the inherent resource restrictions associated with SMEs. Finally, human well-being requires the equitable distribution of both the benefits and the costs of economic restructuring (European Environment Agency, 2013). Therefore, increasing environmental concerns necessitate a large-scale governmental and corporate economic response, one which ideally entails collaborative action by relevant stakeholders in each context.

1.2 Defining the green economy

Before proceeding, it is a worthwhile exercise to contemplate definitions in context. Despite increased research, policy and practitioner attention resulting in a growing body of literature and research spanning multiple disciplines and sectors, there appears to be no generally accepted international definition of a green economy. Previous attempts to define the green economy and green growth cover a spectrum of different 'shades of green', from narrow concerns about climate change on the one hand (in line with initial approaches to low-carbon growth), to larger critiques of the environmental sustainability of modern capitalism on the other

(Green Growth Leaders, 2011). Otherwise stated, from a scientific perspective, the focus varies between addressing one of the planetary boundaries defined by scientists (that is, climate change) to addressing more holistically the wider range of resource and ecological limits and the general state of the planet. As the European Environment Agency (2013) articulates, 'the term "green economy" is not consistently defined, as it is still an emerging concept'.

In truth, the terms 'green economy' and 'green growth' are often used interchangeably despite the fact that green growth occurs from the bottom up, at an operational or a process level, whilst the green economy occurs from the top down, at a macro or strategic level (International Chamber of Commerce, 2012, p. 10). Moreover, there is no set of guiding principles on what constitutes the green economy or how government and policy can truly achieve it. Thus, while there is a general consensus on theme, that is, protecting our environment and natural resources through increased consciousness, there is still an overall lack of definitional scope and precision, leading to a misuse of terminology that ignores the real meaning of green in the context of economic, environmental and social sustainable development.

Prior studies have typically emanated from a variety of disciplines and perspectives, which can be partly attributed to the current confusion in the literature. For example, definitions include 'a resilient economy that provides a better quality of life for all within the ecological limits of the planet' (Green Economy Coalition, 2012) versus 'an economy in which economic growth and environmental responsibility work together in a mutually reinforcing fashion while supporting progress on social development' (International Chamber of Commerce, 2012). Thus, the term has come to be used rather loosely to refer to a broad range of topics from clean tech and renewables, to green business, job creation and public policy, and the protection of the planet, often depending on context and author. The challenge is therefore to amalgamate existing theories and definitions as far as possible to allow some level of dialogue and shared understanding to occur within this text. Therefore, while the authors acknowledge the limitations of a single definition in light of the above, they have adopted the most widely accepted definition as the following:

> A green economy is one that results in improved human well-being and social equity, while significantly reducing environmental risks and ecological scarcities. In its simplest expression, a green economy is low-carbon, resource-efficient, and socially inclusive. (UNEP, 2011, p. 16)

This definition positions the green economy from an ecological economics perspective as being a low-carbon, resource-efficient and socially inclusive economy that enhances energy and resource efficiency, and prevents the loss of biodiversity and ecosystem services. If we accept that 'the concept of green economy focuses primarily on the intersection between environment and economy' (The United Nations, 2010) and that 'achieving sustainability rests almost entirely on getting the economy right' (UNEP, 2011, p. 17), then policy and practice should ideally operate in unison to ensure the green economy is embraced, and contributed to, by regional economies.

1.3 The policy gap

The European Commission (EC) (2007) has established objectives, targets and measures for energy efficiency and renewable energy sources (RES) under its energy policy. Specific measures include

- a 20 per cent reduction in total EU greenhouse gas emissions (including both energy and non-energy-related sources) from 1990 levels;
- a 20 per cent increase in energy efficiency by 2020; and
- an increase in the share of EU energy consumption produced from renewable resources to 20 per cent.

The greenhouse gas and renewable energy targets (EC, 2007) were translated into legislation through the Climate and Energy Package (EC, 2009), while the energy efficiency objective was translated by the Energy Efficiency Plan (EC, 2011) into a 20 per cent reduction in consumption of primary energy compared to energy consumption projections for that year.

In addition to EU policy requirements, the OECD and others recognise green growth as an important contributor to Europe's future prosperity. The OECD Green Growth Strategy report, 'Towards Green Growth' (2011b), aims to provide a practical framework for governments to simultaneously boost economic growth and protect the environment. These frameworks call for a longer time horizon in economic policy decisions, eliminating the short- or medium-term mindset previously adopted. The OECD (2011b) specifically mentions the significance of 'green innovation' as a catalyst for a green growth strategy, as it reduces dependence on established ways of doing things and can 'help to decouple growth from natural capital depletion' (p. 10). However, changing the way in which society manages the interaction of environmental and economic domains requires actions across all sectors (European Environment Agency, 2013), and there has been only limited success in engaging regional SMEs in the green movement to date.

Despite the EC (2007) energy policy, the OECD Working Paper Series (2013)[1] cautions that further policy changes are needed to ensure that both developed and developing countries move to a 'new green growth path' to protect our environment and natural resources. The studied regions offer a number of insights into the green policy-practice interplay within Europe in this regard.

For example, in Ireland, at national level, the drive for green growth is strongly underpinned by the view that green innovation must prevail in environmentally oriented organisations to realise Ireland's green economy ambitions (Comhar Sustainable Development Council,[2] 2009; Government of Ireland, 2012; Government of Ireland, 2011–2014; National Recovery Programme 2011–2016). Comhar's Green New Deal (GND, 2009) represents an important contribution and asserts that central to any sustainable development is the replacement of fossil-fuel energy production with renewable energy, while a key strategy is to 'promote the green enterprise sector and the creation of "green-collar" jobs' (Government of Ireland Report, 2008, p. 7). However, despite numerous policy developments coupled with the establishment of key institutions for environmental protection and sustainable development, including the Environmental Protection Agency in 1993, and the Comhar Sustainable Development Council in 1999, each of which has made significant contributions to green growth strategic planning, these policies have not been fully translated into widespread regional green initiatives to date.

Wales is one of only three nations to have the concept of sustainable development at the very heart of government policy. The Well-being of Future Generations [Wales] bill (Welsh Government, 2014a) strengthens Wales' commitment to sustainable development, through binding goals and indicators, including a legislative programme which 'provides new powers, duties and institutional capacity to advance our goals of building a sustainable Wales' (Welsh Government, 2011). Notably, renewable energy capacity grew by over a third between 2007 and 2010 in the studied regions of Wales. This suggests that Wales' environment policy coupled with small but well-connected regional communities may be well placed to take advantage of the developing green economy. Although the Welsh government is very positive about progress made in the last ten years, in reality a balance between economy and ecology has yet to be achieved. Thus, while Gross Value Added (GVA) per head has gone up, biodiversity and social justice have remained worryingly flat (Welsh Government, 2014b).

Despite the introduction of considerable policy changes covering a wide spectrum of the economy, the actual implementation of a sustainable development policy has not been as successful as originally anticipated in either domain. These findings are reinforced at the European level, where research analysis reflects mixed performance results. Europe has made more progress in improving resource efficiency than preserving ecosystem resilience (Environmental Indicator Report, European Energy Agency, 2012), suggesting a residual preference for 'the prevailing model of economic growth' (European Environment Agency, 2013, p. 5).

This anomaly between policy and practice may be partly due to confusion as to what 'sustainable development' means amongst government and policymakers around the globe (Colby, 1991), with no one agreed definition despite the many proposed. It may also reflect the challenges of translating a complex strategic vision into concrete and measurable goals, targets and indicators, supported with appropriate communication, participation, continuing assessment and institutional capacity (Bellagio principles on assessing sustainable development, International Institute for Sustainable Development, IISD, 1997). Common to all understandings, however, is the view that the green economy is a key driver of future growth, economic recovery and development. Thus, the policy gap is starting to close, with a greener mindset that recognises sustainable development will not happen by itself.

When contemplating the green economy transition, the authors calibrated the component criteria associated with the translation of the incumbent strategic vision (Figure 1.1). Here, the stakeholders are identified as government, policymakers, business and society/community. The primary focus of this book is on regional SME engagement with the green economy. The capabilities required to facilitate a green economy transition in regional SMEs are identified as a long-term and forward-oriented business focus, an outward-looking sustainability planning ethos, and concentration on innovation capability enhancement that promotes

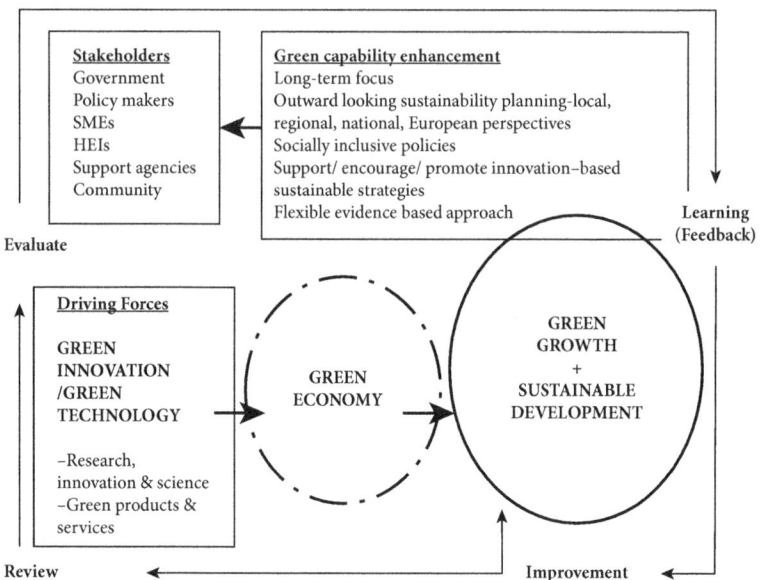

FIGURE 1.1 *Green economy transition in regional SMEs*
Source: Adapted by author(s) from [Current research]

socially inclusive and collective commercial activities within the context of appropriately communicated supports to ultimately stimulate green growth and sustainable development while protecting the environment.

SMEs have been considered one of the 'driving forces' and the 'backbone' of modern economies due to their contributions in terms of technological upgrading, product and process innovations, employment generation and export promotion (European Commission, 2012). More specifically, SMEs actively participate in the sectors that are the focus of recent green-oriented policies, such as renewable energy production, smart metering, building refurbishment, cleaner cars, wind and solar installations, and battery development (International Energy Agency, IEA, 2009). The creativity and dynamism displayed by SMEs means they have a crucial role to play in this area – both as eco-innovators and as recipients of green technologies (European Commission, 2010). It is therefore crucial to better understand and convey the role of SMEs in the green economy and particularly, how SME knowledge intensity levels, together with their research and development, and innovation practices, can have a significant impact on building a green economy.

1.4 The green business landscape in Ireland and Wales – a regional perspective

The statistics relating to SMEs in the studied regions emulate those of Europe: 99.8 per cent of Irish firms are SMEs (Irish SME Association, 2014),[3] while SMEs make up 99.9 per cent of private sector businesses in Wales (Federation of Small Businesses, 2014).[4] Recently, the EC published the results of a detailed survey that examined the views of over 11,000 SMEs in the 28 EU member states towards key environmental issues (European Commission, 2013). The survey found that, across Europe, SMEs are creating green jobs, improving their resource efficiency and making a significant contribution to a low-carbon economy. Cost savings seem to be the primary reason for this changing landscape, with the research finding that the majority of SMEs in the EU act to become more resource-efficient in order to reduce costs (63 per cent), although 28 per cent say the environment is one of the top priorities for their enterprise. From a policy viewpoint, the survey showed that greater support by means of grants and subsidies could help SMEs become more efficient and, more importantly, introduce green goods and services in pursuit of enhanced effectiveness.

Across Wales, there are now more low-carbon and environmental jobs (41,506) than in motor trades (21,500), financial services (27,800) and telecommunications (6,600). The outstanding terrestrial and marine environment of Wales attracts high-level expertise to its universities and public bodies, including national and local government and agencies, as people with interests in and commitment to sustainability, or 'green issues', are attracted and retained through lifestyle and career choices. Within the environmental goods and services (EGS) sector, the three domains with the highest sales in 2010–11 were building technologies (£817m), water supply and waste water treatment (£628m), and geothermal (£637m).

According to the Government's Progress Report on Green Growth and Employment, Ireland has made significant progress in the areas of water, waste management, renewable energy, bioenergy and green tourism, as well as research and development across eco-innovation and green technology (Government of Ireland, 2013). Ireland also advanced the implementation of green initiatives and employment in international sectors, including renewable energy, smart grids, sustainable food, tourism and energy-efficient products and services (OECD, 2014). The country's GNP goal coincides with OECD recommendations that Government adopt a medium- to long-term focus for advancing sustainable economic and human development and engineering the green economy. Furthermore, based on recent survey results (European Commission, 2013), Ireland has a full-time green employee in 42 per cent of its SMEs, while the value of the Irish EGS market, which is dominated by domestic SMEs, is conservatively estimated at €2.8 billion (Forfás in collaboration with InterTradeIreland, 2008).

Based on current experience in the observed regions' SMEs and the reported insights at European level (Environmental Indicator Report, 2012), Europe has made more progress in improving resource efficiency than preserving ecosystem resilience. Thus, while improvements in resource efficiency are keeping pace with the increases in consumption, real structural change calls for a broad transformation across markets, supply chains, resource efficiency and recirculation, infrastructure, organisations and consumption patterns. This effort needs to be driven by a cross-disciplinary innovativeness culture and willingness to change and adapt established ways of working. Indeed, this transition requires a unique set of capabilities and resources that are developed over time to enable green growth and sustainable development for enhanced competitiveness and survival of the economy, environment and society.

Notes

1 http://www.oecd.org/industry/publicationsdocuments/workingpapers/.
2 The forum for national consultation and dialogue on all issues surrounding Ireland's pursuit of sustainable development. In January 2012, the sustainable development role performed by Comhar, the Sustainable Development Council (SDC), was integrated into the work of the National Economic and Social Council (NESC). NESC has since developed its work in a way that integrates sustainable development issues into its analysis of significant national challenges.
3 http://isme.ie/.
4 http://www.fsb.org.uk/stats.

References

Colby, M. 1991. Environmental management in development: the evolution of paradigms. *Ecological Economics*, 3, 193–213.

Comhar Sustainable Development Council. 2009. *Towards a Green New Deal for Ireland*. [Internet] Available at: http://files.nesc.ie/comhar_archive/Comharper cent20Reports/Comhar_25_2009.pdf [Accessed 2 April 2014].

Ernst and Young, 2012. *Cleantech Ireland: An Assessment of the Sector and the Impact on the National Economy*. Ireland: Ernest and Young in conjunction with Oxford Economics.

European Commission. 2007. *Renewable Energy Road Map – Renewable Energies in the 21st Century: Building a More Sustainable Future*. [Internet] Available at: http://ec.europa.eu/energy/energy_policy/doc/03_renewable_energy_roadmap_en.pdf [Accessed 2 April 2014].

European Commission. 2010. *Making Eco-Innovation Happen in Small and Medium-sized Enterprises*. 8th European Forum on Eco-Innovation, 20–21 April 2010, Bilbao, Spain. [Internet] Available at: http://ec.europa.eu/environment/archives/ecoinnovation2010/1st_forum/pdf/case_studies.pdf [Accessed 12 February 2014].

European Commission. 2011. *Energy Efficiency Plan*. [Internet] Available at: http://ec.europa.eu/energy/efficiency/action_plan/action_plan_en.htm [Accessed 2 April 2014].

European Commission. 2012. *EU SMEs in 2012: At the Crossroads – Annual Report on Small and Medium-sized Enterprises in the EU, 2011/12*. Rotterdam, September 2012. [Internet] Available at: http://ec.europa.eu/enterprise/policies/sme/facts-figures-analysis/performance-review/files/supporting-documents/2012/annual-report_en.pdf [Accessed 2 April 2014].

European Commission. 2013. *SMEs, Resource Efficiency and Green Markets*. [Internet] Available at: http://ec.europa.eu/public_opinion/flash/fl_381_en.pdf [Accessed 2 April 2014].

European Energy Agency. 2012. *Environmental Indicator Report 2012 – Ecosystem Resilience and Resource Efficiency in a Green Economy in Europe*. EEA, Copenhagen. Pages: 151, ISBN 978–92-9213–315-3.

European Environment Agency. 2013. *Towards a Green Economy in Europe: EU Environmental Policy Targets and Objectives 2010–2050*. Luxembourg: Publications Office of the European Union, EEA Report No. 8/2013.

Forfás and InterTradeIreland. 2008. *Environmental Goods and Services Sector on the Island of Ireland – Enterprise Opportunities and Policy Implications*. [Internet] Available at: http://www.intertradeireland.com/media/Environmentalper cent20Goodsper cent20andper cent2oServicesper cent2oSectorper cent2oonper cent2otheper cent2oIslandper cent2oofper cent2oIrelandper cent2oFinalper cent2oReport.pdf [Accessed 12 February 2014].

Forfás. 2011. *Progress Report on the Implementation of the Recommendations of the Report of the High Level Group on Green Enterprise.* [Internet] Available at: http://www.forfas.ie/media/forfas-110318-green_enterprise. pdf [Accessed 2 April 2014].

Government of Ireland. 2008. *Building Ireland's Smart Economy – A Framework for Sustainable Economic Renewal.* ISBN 978-1-4064-2244-3, [Internet]. Available at: http://www.taoiseach.gov.ie/attached_files/BuildingIrelandsSmartEconomy.pdf [Accessed 2 April 2014].

Government of Ireland. 2011. *National Recovery Plan 2011-2014.* [Internet] Available at: http://www.budget.gov.ie/Theper cent2oNationalper cent2oRecoveryper cent2oPlanper cent202011-2014.pdf [Accessed 12 February 2014].

Government of Ireland. 2012. *Delivering Our Green Potential – Government Policy Statement on Growth and Employment in the Green Economy.* [Internet] Available at: https://www.agriculture.gov.ie/media/migration/ruralenvironment/environment/bioenergyscheme/DeliveringOurGreenPotential171212.pdf [Accessed 2 April 2014].

Government of Ireland. 2013. *Progress Report on Green Growth and Employment in the Green Economy in Ireland.* [Internet] Available at: http://www.djei.ie/publications/enterprise/2013/Green_Economy_Progress_Report_2013.pdf [Accessed 30 January 2014].

Green Economy Coalition. 2012. *Submission to UNCSD Zero Draft Text.* [Internet] Available at: http://www.greeneconomycoalition.org/sites/greeneconomycoalition.org/files/documents/Greenper cent2oEconomyper cent2oCoalitionper cent2oZeroper cent2oDraftper cent2otextper cent2oper cent28finalper cent29_0.pdf [Accessed 30 January 2014].

Green Growth Leaders. 2011. *Shaping the Green Growth Economy – A Review of the Public Debate and the Prospects for Green Growth.* EEA, Copenhagen. [Internet] Available at: http://www.sustainia.me/resources/publications/mm/Shaping-the-Green-Growth-Economy_report.pdf [Accessed 30 January 2014].

International Chamber of Commerce. 2012. *Green Economy Roadmap – A Guide for Business, Policymakers and Society.* [Internet] Available at:

http://www.iccwbo.org/products-and-services/trade-facilitation/green-economy-roadmap/ [Accessed 12 February 2014].

International Energy Agency, IEA. 2009. *World Energy Outlook*, IEA, Paris, [Internet] Available at: http://www.worldenergyoutlook.org/media/weowebsite/2009/WEO2009.pdf [Accessed 2 April 2014].

International Institute for Sustainable Development, IISD. 1997. *Assessing Sustainable Development – Principles in Practice.* [Internet] Available at: http://www.iisd.org/pdf/bellagio.pdf [Accessed 2 April 2014].

IPCC. 2014. Summary for policymakers. O. Edenhofer, R. Pichs-Madruga, Y. Sokona, E. Farahani, S. Kadner, K. Seyboth, A. Adler, I. Baum, S. Brunner, P. Eickemeier, B. Kriemann, J. Savolainen, S. Schlömer, C. von Stechow, T. Zwickel and J.C. Minx (eds) *Climate Change 2014, Mitigation of Climate Change: Contribution of Working Group III to the Fifth Assessment Report of the Intergovernmental Panel on Climate*, Cambridge, UK, and New York: Cambridge University Press.

OECD. 2011a. *Fostering Innovation for Green Growth*, Pages: 130, ISBN 978–92-64–11991-8. [Internet] Available at: http://www.keepeek.com/Digital-Asset-Management/oecd/science-and-technology/fostering-innovation-for-green-growth_9789264119925-en#page9 [Accessed 20 March 2014].

OECD. 2011b. *Towards Green Growth,* Number of pages: 142 ISBN: 9789264094970. [Internet] Available at: http://www.oecd.org/greengrowth/48224539.pdf [Accessed 20 March 2014].

OECD. 2012. *OECD Environmental Outlook to 2050 – The Consequences of Inaction.* OECD Publishing. ISBN 978–92-64–12216-1 (print), ISBN 978–92-64–12224-6 (PDF) [Accessed 20 March 2014].

OECD. 2013. *Working Paper Series.* [Internet] Available at: http://www.oecd.org/industry/publicationsdocuments/workingpapers/ [Accessed 20 March 2014].

OECD. 2014. *OECD Environmental Performance Review of Ireland – Mid-term Progress Report. Paris,* 26–28 March. [Internet] Available at: http://www.oecd.org/env/country-reviews/Ireland_EPR_MidTermReport.pdf [Accessed 2 April 2014].

United Nations, 2010. *The Millennium Development Goals Report.* New York: MDG Report 2010 En 20100604 r14 Final.indd 1.

United Nations Department of Economic and Social Affairs (UNDESA) Division for Sustainable Development. 2012. *A Guidebook to the Green Economy Issue 1: Green Economy, Green Growth, and Low-Carbon*

Development – History, Definitions and a Guide to Recent Publications. Prepared by Cameron Allen and Stuart Clouth. August 2012. [Internet] Available at: http://sustainabledevelopment.un.org/content/documents/GEper cent20Guidebook.pdf [Accessed 12 January 2014].

United Nations Environment Programme (UNEP). 2011. *Towards a Green Economy: Pathways to Sustainable Development and Poverty Eradication – A Synthesis for Policy Makers.*[Internet] Available at: http://www.unep.org/greeneconomy/Portals/88/documents/ger/ger_final_dec_2011/Green%20EconomyReport_Final_Dec2011.pdf [Accessed 12 January 2014].

United Nations Environment Programme, UNEP. 2014. *Green Economy Initiative.* [Internet] Available at: http://www.unep.org/greeneconomy/aboutgei/whatisgei/tabid/29784/default.aspx [Accessed 20 March 2014].

Welsh Government. 2011. *The Welsh Government's Legislative Programme 2011–16.* [Internet] Available at: http://wales.gov.uk/legislation/programme/?lang=en [Accessed 7 February 2014].

Welsh Government. 2014a. Well-being of Future Generations [Wales] bill. [Internet] Available at: http://wales.gov.uk/legislation/programme/assemblybills/future-generations/?lang=en [Accessed 30 January 2014].

Welsh Government. 2014b. *One Wales One Planet – The Sustainable Development Annual Report 2013–14.* [Internet] Available at: http://www.assemblywales.org/bus-home/bus-business-fourth-assembly-laid-docs/gen-ld9791-e.pdf?langoption=3&ttl=GEN-LD9791per cent20-per cent20Oneper cent20Walesper cent3Aper cent20Oneper cent20Planetper cent20-per cent20Theper cent20Sustainableper cent20Developmentper cent20Annualper cent20Reportper cent202013–14 [Accessed 20 February 2014].

2
The Green Innovation and Future Technologies (GIFT) Concept

L. Reinl and F. Kelliher

Abstract: *The GIFT concept hinges on integrated stakeholder engagement between government agencies, HEIs, SMEs, economic support groups and rural development groups, underpinned by a learning community philosophy. Once conceived, GIFT gathered cross-country/regional HEI colleagues from the natural, social and physical sciences to pursue the design, development and delivery of a collaborative green innovativeness education programme in interaction with the wider regional community. The chapter documents findings around how to leverage innovation in pursuit of green capability development within regions, and concludes with insights into how cross-border innovation and knowledge exchange happen in a green learning community. Finally, the authors propose a framework of regional SME engagement for sustainable development.*

Kelliher, Felicity and Leana Reinl. *Green Innovation and Future Technology: Engaging Regional SMEs in the Green Economy.* Basingstoke: Palgrave Macmillan, 2015.
DOI: 10.1057/9781137479822.0007.

As a first step in the GIFT journey, HEI colleagues from the natural, social and physical sciences in each of the participant regions formed an interdisciplinary academic team. The goal was to pursue the design, development and delivery of a collaborative green innovativeness education programme in interaction with the wider regional business community. As outlined in the introduction, the success of this goal hinged on integrated stakeholder engagement between government agencies, HEIs, indigenous businesses, economic support groups and rural development groups (Döring and Schnellenbach, 2006; Drda-Kühn and Wiegand, 2010; Kelliher et al., 2014). This type of regional engagement represents a structured approach to SME development that is underpinned by a learning network/community philosophy (Bessant and Tsekouras, 2001; Reinl and Kelliher, 2010) and informed by research promoting regional learning and multilayered knowledge exchange in pursuit of innovation (Döring and Schnellenbach, 2006; Haugen Gausdal, 2008). The underlying assumption is that innovative activity develops within such learning communities (Mitra, 2000); however, this development is 'not to be seen as a simple and spontaneous event' (Novelli et al., 2006, p. 1150).

This chapter considers the conditions which facilitate green innovativeness in a multilevel regional SME stakeholder context. As SMEs face very different resource challenges to their larger counterparts (Kaufmann and Tödtling, 2002; Kelliher and Reinl, 2009), particularly in relation to the development of a green innovativeness skill set (Santoro and Chakrabarti, 2002; Walsh et al., 2012), they tend to innovate within a set of interactive relationships (Aylward, 2012; Reinl and Kelliher, 2010). Before contemplating a framework to impart green innovation skills to regional SMEs, we consider the conditions which facilitate green innovativeness and green capability development in an SME context.

2.1 Green innovativeness and green capability development – An SME perspective

Envisioning how to leverage innovation in pursuit of green capability development, we note that skill development through the leveraged innovation resource of firms (Walsh et al., 2012) leads to an underlying capability structure and cultural willingness, which in turn enables the firm to produce eco-friendly innovations. Such an innovation-based sustainable development strategy (Mariadoss, Tansuhaj and Mouri, 2011) should be driven by the enhancement of small firm internal capabilities, which should precede the development of sustainability-based managerial practices (Kucera, 2009; Reinl and Kelliher, 2010).

Drawing from the resource-based view and dynamic capabilities theory (Barney, 1991; Eisenhardt and Martin, 2000), we contend that a firm's 'innovation capability' is composed of a technological and behavioural dimension, denoting both technological capacity and behavioural willingness and commitment of the firm to innovate (Walsh et al., 2012). This in turn influences the development of innovation-based sustainable strategies and enhances the SME's green capabilities, offering insights into the potential for HEIs to contribute to this skill development in a meaningful way.

The Green Business Innovation Capabilities (GBICs) literature (Lin, Tseng, Chen and Chui, 2010; Walsh et al., 2012) suggests that organisations must strategically and routinely reorganise their innovation capabilities to harmonise innovation with the external competitive environment and wider society. If one accepts that research, development, training and innovation are multifaceted and essentially interlinked, there are ways of adding high-level value, including new product development, product modification, new process development and process modification. Each of these value-enhancing interventions depends on high calibre management, skilled in each realm, in order to succeed (Simango, 2000). Building on Simango's (2000) perspective, GBICs will allow SMEs to evaluate the required technical prowess, knowledge and skills, as well as recognise the economic and commercial feasibility of green innovativeness in order to reduce the risk of innovation activities, since innovation by its very nature is a high-risk action (Fell, Hansen and Becker, 2003).

Assuming there is a positive relationship between a firm's level of innovativeness and its willingness to change and accept new ways of doing things and invest greater efforts in research and development, an SME's level of innovativeness may be a key predictor of its likelihood to engage in green innovation training initiatives and/or projects in response to environmental trends and changes. Subsequently this may result in a sustainable competitive advantage, but only if supported by interdisciplinary collaborative action on the part of regional stakeholders (Aylward, 2012; Murphy, 1993) as comprehensive stakeholder engagement is a vital component when pursing transformational innovative capability (Dutta et al., 2005; Lado and Wilson, 1994; Wang and Ahmed, 2007) in these firms.

Considering the above, and given the underlying principles of resource mapping and limitations, modelling and management, multidisciplinarity and inclusivity discussed previously, a structured approach to knowledge exchange is endorsed in this context. Here, collaborative relationships have the potential to enhance SME innovation capabilities (Aylward, 2012; Murphy, 1993), functioning as a catalyst of change in pursuit of green economic development. This enhancement is achieved by engaging stakeholders in multigeographic virtual collaborative learning spaces[1] (Lin et al., 2010), wherein knowledge brokers facilitate the sharing of experience and the diffusion of ideas (Jørgensen and Keller, 2008; Reinl and Kelliher, 2014) among stakeholders within learning sets.[2]

This collegial process of knowledge exchange challenges the existing frames of reference of SME owner/managers, and in turn they develop an appreciation of the interdependencies of sustainability issues and requirements for collective action in their resolution (Graci, 2013, p. 36). Through continued interaction a repertoire of stories, rules and routines permits stakeholders to engage in shared practice (Swan, Scarbrough and Robertson, 2002; Wenger, 1998), and in doing so, to know something new. As such flows of knowledge are 'inextricably linked to the social relations which develop through shared practice' (Swan, Scarbrough and Robertson, 2002, p. 479). Socially constructed sets of relationships (Johannisson, 1995; Lave and Wenger, 1991) permit stakeholders 'sharing a concern or set of problems' to develop knowledge and expertise through interaction with one another on ongoing basis (Wenger, McDermott and Snyder, 2002) in a continuous community of practice cycle (Wenger, 1998).

2.2 Embedding cross-border innovation and knowledge exchange in a green learning community

As previously discussed SMEs do not innovate on their own but within a set of interactive relationships (Aylward, 2012; Reinl and Kelliher, 2010) where collaborative engagement can 'help foster an environment in which knowledge can be created and shared and, most importantly, used to improve effectiveness, efficiency, and innovation' (Lesser and Everest, 2001,

p. 46). Thus, collaborative stakeholder engagement is a vital component when pursing transformational innovative capability (Dutta et al., 2005; Lado and Wilson, 1994; Wang and Ahmed, 2007) in regional SMEs. The GIFT framework as represented in Figure 2.1 permits shared practice to evolve through collaborative and comprehensive stakeholder engagement, activities that are crucial in the pursuit of transformational innovative capability (Dutta et al., 2005; Lado and Wilson, 1994; Wang and Ahmed, 2007) in regional SMEs. The proposed framework recognises that the structures which facilitate and support SME engagement must function at a competent level (Johannisson, 2007) to sustain knowledge-exchange value in the longer term, and that these structures take time to develop. This premise dictates a phased approach to individual and collective engagement and green capability development in which identities and practice can emerge and develop over time through the dynamics of voluntary, trusting relationships (Miles and Tully, 2007; Reinl and Kelliher, 2010). In time, and with appropriate broker[3] support (Burt, 2005; Gulati and Garguilo, 1999; Kelliher and Reinl, 2009), the sharing of knowledge, expertise and physical resources permits stakeholders to exploit shared resources, which in turn enable the creation of new strategic paths to future competitiveness.

It is clear from the above that while a significant pool of knowledge may be present in and among regional SMEs and stakeholders, external catalysts are often required to trigger SME innovation (Reinl and Kelliher, 2010; Lundberg and Tell, 1998). Given that SME engagement is framed by a restricted resource base (Kaufmann and Tödtling, 2002; Kelliher and Reinl, 2009), these firms tend to suffer knowledge-exchange deficits due to an over-reliance on local networks (Freel, 1999; Saxena and Ilberry, 2008; Wickham, 2001), which can create boundaries within and across regions (Aylward, 2012). Therefore ideal learning spaces should promote broader stakeholder inclusion on both a continuous and an interim basis, open knowledge-exchange boundaries (Jack, Anderson, and Dodd, 2004; Reinl and Kelliher, 2014) to provide an important outside-in perspective, instigate wider reflection (Kelliher and Reinl, 2011; Wenger, 1998) and counteract the potential of the learning community's becoming insular as time passes (Wenger, 1998). This should enable resource release, as objects of learning[4] may traverse the boundaries of the community of practice (Kelliher and Reinl, 2011; Wenger, 1998).

Brokers seek to support 'the creation and further development of a common base of shared knowledge among individuals...to co-ordinate their actions in the resolution of the technological and organisational problems that they confront' (Keeble, 2000, in Haugen Gausdal, 2008). They endeavour to broaden knowledge-exchange boundaries to include international actors (Halme, 2001) as a means to alleviate 'spatial blindness' (Brouder and Eriksson, 2013) and release knowledge flows. However, when the learning community boundary spans the distance between countries, additional challenges and barriers to embedding innovation and knowledge exchange arise and require consideration.

While in situ knowledge brokers (Phillipson, Gorton and Laschewski, 2006; Reinl and Kelliher, 2014) representing key green subsectors are well placed to readily 'translate' local knowledge (Breschi and Lissoni, 2001), the GIFT team also sought to gain access to and interaction with an international knowledge base. Specifically, the GIFT virtual and physical 'collaborative space' (Zhu, 2012), embedded in online discussion boards, facilitates broader access to emergent knowledge for geographically distant regional stakeholders. Through the resultant virtual learning environment (VLE), knowledge brokers, in interaction with discipline/topic experts, continually move knowledge and ideas in and out of the regional network, facilitating the exchange of experiences in an international context (Halme, 2001). Notably, the VLE in this project permits cyclical engagement with core GIFT themes (green tourism, green technology, the knowledge economy and waste management), and in doing so, create the 'social proximity' required to underpin shared practice (Lave and Wenger, 1991; Miles and Tully, 2007; Reinl and Kelliher, 2010) and promote knowledge flows (Breschi and Lissoni, 2009; Lave and Wenger, 1991) in a cross-border learning community.

In the latter section of this chapter, a framework of regional SME engagement for sustainable development is proffered. The framework illustrates the process through which the above conditions are sought and ultimately how multilevel, cross-border collaborative action is built.

2.3 Proposing a framework of regional SME engagement for sustainable development

As noted previously, the authors recognised that the pursuit of skill enhancement needed to be based on comprehensive collaborative action within an academic-practice partnership ethos (Kelliher, Harrington and Galavan, 2010). As such, the initial GIFT focus was to promote stakeholder engagement and interaction via physical and virtual interactions, with specific green-based topics offering a basis for shared practice that can develop and in turn sustain more complex cross-border knowledge exchanges between participants. Once trust was established, the GIFT team sought to leverage that engagement via business-led academe-facilitated capability enhancement programmes, ultimately incorporating more complex collaborative business-led green projects over time. This phased approach sought to solidify a knowledge-reciprocity cycle in the hope that ongoing cyclical engagement in physical and virtual fora might potentially underpin shared practice and build the trust required to support a sustainable community of practice (Lave and Wenger, 1991; Miles and Tully, 2007; Reinl and Kelliher, 2010). Here, a VLE may provide a linchpin to embed a sustainable cross-border CoP after capability enhancement programmes come to an end.

The generation and integration of regional green innovativeness required an ethos of knowledge integration in pursuit of skill enhancement (Kelliher et al., 2010) at local, national and international level through a collaborative interdisciplinary cross-border approach which sought to

- develop a cross-border green knowledge/innovation 'ecosystem' comprising regional stakeholders to facilitate the green agenda through a portfolio of studies and actions;
- provide technically based training activities relating to business, the environment, climate change, energy and waste management;
- facilitate greater linkages within and between the various disciplines in HEIs and regional industry; and
- apply business activities which utilise the environment and can grow through green ways of working and new eco-products/ services.

This evolutionary process is exemplified in the above four-stage approach: initial regional network engagement, a focused professional development programme, cross-border collaborative space facilitated through physical and virtual fora, and in-depth academic-SME collaborative projects. It should be noted that these stages are not necessarily chronological and dependent on stakeholder requirements and the culture of a particular region. Some may be enacted simultaneously, and some may operate beyond the programme duration. Regardless, this approach is fundamental to the building of successful academic-business relationships which can be expanded to reciprocal innovation knowledge transfer in a cross-border area. These criteria form the basis from which a CoP framework of cross-border stakeholder engagement is now proposed (Figure 2.1).

Once the regional SME community was engaged in the GIFT programme in each country through a process of initial identification, individual contact and a series of green workshops and events, followed by interspersed expert sessions (with for example, ecotourism specialists, regional funding authorities, biofuel engineers and wind energy experts), a VLE provided the linchpin to embedding this cross-border community of practice in the green ethos. Here, business and academic experts across the disciplines existed in a virtual space alongside SME owners on a bimonthly basis, and through cyclical engagement with specific green-based topics, all parties could contemplate initial consideration relating to business challenges and opportunities. Over time, this cyclical process helped identify the core GIFT themes (Green Tourism, Green Technology, the Knowledge Economy and Waste Management) in interaction with each regional SME community. These themes were

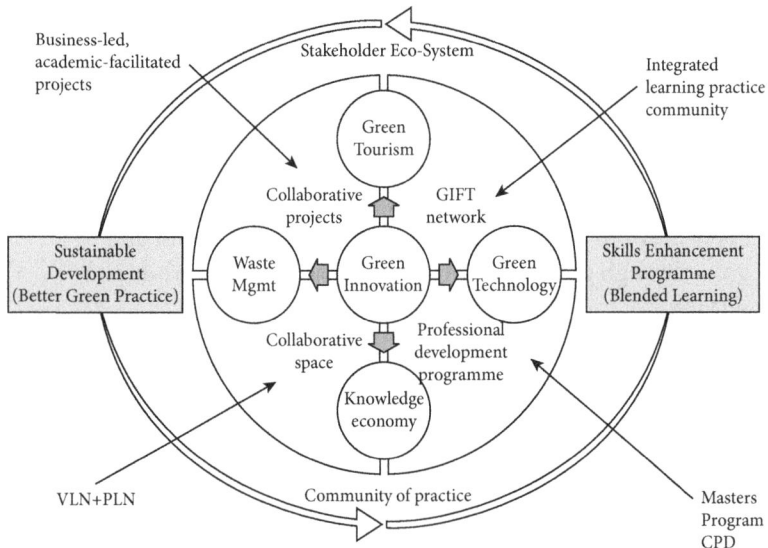

FIGURE 2.1   *CoP framework for cross-border stakeholder engagement*
Source: adapted from Kelliher et al. (2013).

found to encompass stakeholder needs in each jurisdiction, and the view held was that these themes could be conceptualised as key determinants of green innovativeness. Here, each theme acts as an antecedent to green innovation as the central mediating action, and in return, green innovation feeds back out into a continuous community of practice cycle (Wenger, 1998), facilitating the emergence of a stakeholder ecosystem (as illustrated in Figure 2.1).

A facilitated contributory-reciprocal cycle (Koch, Kautonen and Grünhagen, 2006) is employed in Figure 2.1, in which green innovation (innovativeness) is set as the central mediating action engaging each of the core GIFT themes. Professional development programmes, such as training in wood fuel quality processes (as detailed in case C), biofuel heating systems and waste management sought to enhance green technology capabilities in participating SMEs, facilitating the production and development of green innovation capabilities in these firms. The framework seeks to use the 'collaborative space' (in this case the virtual fora) to identify collegial projects of mutual benefit to engaged regional SMEs. Subsequently, regional HEIs sought to conduct these (small) collegial research projects, carried out by HEI students in the SME under the mentorship of experienced academics, which result in 'new ideas, devices or methods' that can be exploited in the short term by the project recipient and in the longer term through knowledge sharing via the GIFT collaborative space. A series of these projects is detailed in Chapter 4 (section 4.1). The GIFT programme also supports undergraduate and postgraduate participation in larger SME/industry-led collaborative projects to facilitate the reciprocal knowledge-transfer activities (as highlighted in Figure 2.1). The presented cases demonstrate how cross-disciplinary engagement may enhance graduate awareness, ability and action in the sustainable development domain, just as Dawe et al. (2005) suggested.

In response to significant student demand for further sustainable development in the course curriculum, GIFT sought to embed sustainable development in the curricula of partner HEIs and in doing so, to expand opportunities for students to access interdisciplinary teaching and international insights. The challenge in developing transformative teaching techniques that can facilitate this cross-disciplinary requirement are discussed in further detail in Chapter 4 (this volume), while the GIFT team's approach to this task is detailed in cases E and F (Chapter 4, this volume).

2.4 Concluding remarks

The particular knowledge exchange challenges that regional SMEs face in relation to green innovation and skill enhancement as outlined in this chapter, informed the framework for cross-border stakeholder engagement presented above (Figure 2.1). The framework illustrates the underlying criteria required to support integrated multilevel stakeholder engagement of Government, HEIs, SMEs, support agencies and regional community groups. Cyclical engagement within a community of practice framework functions as a catalyst of change which can result in transformational SME innovative capability development given appropriate broker support. In Chapters 3 and 4, a series of practical cases further illustrates the multilevel engagement, criteria and concepts discussed in this chapter.

Notes

1. A term adopted to describe the physical and virtual fora in the GIFT programme and framework (Lin et al., 2010).
2. Learning sets comprise SME owner/managers, external experts and experienced facilitators that encourage active participation and experience sharing among regional stakeholders.
3. A broker is an individual or firm that acts as an intermediary between two or more parties (Burt, 2005).
4. In a learning community, learning objects encompass and give form to the experience and understanding of members. Refined within a democratic negotiation domain, these objects permit the learning community to continuously produce new meaning and knowledge (Handley et al., 2006; Wenger, 1998).

References

Aylward, E. 2012. Collaborative Rural Networks (CRNs): An Examination of the Roles and Relationships between Regional Stakeholders, PhD thesis, Waterford Institute of Technology, Ireland.

Barney, J.B. 1991. Firm resources and sustained competitive advantage, *Journal of Management*, 17(1), 99–120.

Bessant, J. and Tsekouras, G. 2001. Developing learning networks, *AI and Society*, 15, 82–89.

Breschi, S. and Lissoni, F. 2001. Knowledge spillovers and local innovation systems: a critical survey, *Industrial and Corporate Change*, 10(4), 975–100.

Breschi, S. and Lissoni, F. 2009. Mobility of skilled workers and co-invention networks: an anatomy of localized knowledge flows, *Journal of Economic Geography*, 9(4), 439–468.

Brouder, P. and Eriksson, R.H. 2013. Staying power: what influences micro-firm survival in tourism? *Tourism Geographies: An International Journal of Tourism Space, Place and Environment*, 15(1), 125–144.

Burt, R.S. 2005. *Brokerage and Closer: An Introduction to Social Capital*, Oxford: Oxford University Press.

Dawe, G., Jucker, R. and Martin, S., 2005. Sustainable Development in Higher Education: Current Practice and Future Developments Higher Education Academy. Available at: www.heacademy. ac.uk/assets/York/ documents/ourwork/tla/sustainability/ sustdevinHEfinalreport.pdf [Accessed 22 June 2013].

Döring, T. and Schnellenbach, J. 2006. What do we know about geographical knowledge spillovers and regional growth? a survey of the literature, *Regional Studies*, 40(3), 375–395.

Drda-Kühn, K. and Wiegand, D. 2010. From culture to cultural economic power: rural regional development in small German communities, *Creative Industries Journal*, 3(1), 89–96.

Dutta S., Narasimhan O. and Rajiv S. 2005. Conceptualizing and measuring capabilities: methodology and empirical application, *Strategic Management Journal*, 26(3), 27–285.

Eisenhardt, K. and Martin, J.A. 2000. Dynamic capabilities: what are they? *Strategic Management Journal*, 21(10/11), 1105–1121.

Fell, D.R., Hansen, E.N. and Becker, B.W. 2003. Measuring innovativeness for the adoption of industrial products. *Industrial Marketing Management*, 32(4), 347–353.

Freel, M.S. 1999. Where are the skills gaps in innovative small firms? *International Journal of Entrepreneurial Behaviour and Research*, 5(3), 144–154.

Graci, S. 2013. Collaboration and partnership development for sustainable tourism, *Tourism Geographies: An International Journal of Tourism Space, Place and Environment*, 15(1), 25–42.

Gulati, R., and Garguilo, M. 1999. Where do inter-organisational networks come from? *American Journal of Sociology*, 104(5), 1439–1493.

Halme, M. 2001. Learning for sustainable development in tourism networks, *Business Strategy and the Environment*, 10(2), 100–114.

Handley, K., Sturdy, A., Fincham, R. and Clark. T. 2006. Within and beyond communities of practice: making sense of learning through participation, identity and practice, *Journal of Management Studies*, 43(3), 641–653.

Haugen Gausdal, A. 2008. Developing regional communities of practice by network reflection: the case of the electronics industry, *Entrepreneurship and Regional Development*, 20(3), 209–235.

Jack, S., Dodd, S. and Anderson, A. 2004. Social structures and entrepreneurial networks: the strength of strong ties, *International Journal of Entrepreneurship and Innovation*, 5(2), 107–120.

Johannisson, B. 1995. Paradigms and entrepreneurial networks: some methodological challenges, *Entrepreneurship and Regional Development*, 7(3), 215–231.

Johannisson, B. 2007. Enacting local economic development – theoretical and methodological challenges, *Journal of Enterprising Communities: People and Places in the Global Economy*, 1(1), 7–26.

Jørgensen, K.M., and Keller, H.D. 2008. The contribution of communities of practice to human resource development: learning as negotiating identity, *Advances in Developing Human Resources*, 10(4), 525–540.

Kaufmann, A. and Tödtling, F. 2002. How effective is innovation support for SMEs? An analysis of the region of upper Austria, *Technovation,* 22(3), 147–159.

Kelliher, F. and Reinl, L. 2009. A resource-based view of micro-firm management practice, *Journal of Small Business and Enterprise Development*, 16(3), 521–532.

Kelliher, F. and Reinl, L. 2011. From facilitated to independent tourism learning networks: connecting the dots. *Journal of Tourism Planning and Development*, 8(2), 185–198.

Kelliher, F., Aylward, E. and Lynch, P. 2014, Exploring rural enterprise: the impact of regional stakeholder engagement on collaborative rural networks. C. Henry & G. McElwee (eds) *Exploring Rural Enterprise: New Perspectives on Research, Policy and Practice*, London: Routledge.

Kelliher, F., Harrington, D. and Galavan, R., 2010, Spreading leader knowledge: investigating a participatory mode of knowledge dissemination among management undergraduates, *Irish Journal of Management*, 29(2), 107–125.

Kelliher, F., Joyner, D., Harrington, D., McDonald, M., Griffiths, G., Owens, E., Walmsley, J. and Walsh, M. 2013, *Institute for Small Business and Entrepreneurship Conference*, 12–13 November, Cardiff.

Koch, L., Kautonen, T. and Grünhagen, M. 2006. Development of cooperation in new venture support networks: the role of key actors, *Journal of Small Business and Enterprise Development*, 13(1), 62–72.

Kucera, D. 2009. Green economy and green jobs: myth or reality? *Sustainable Development: A Challenge for European Research Conference Proceedings*, 26–28 May, Brussels.

Lado, A.A. and Wilson, M.C. 1994. Human resource systems and sustained competitive advantage: a competency-based perspective, *Academy of Management Review*, 19(4), 699–727.

Lave, J. and Wenger, E. 1991. *Situated Learning: Legitimate Peripheral Participation*. Cambridge: University of Cambridge Press.

Lesser, E. and Everest, K. 2001. Using communities of practice to manage intellectual capital, *Ivey Business Journal*, 65(4), 37–41.

Lin, Y., Tseng, M.L., Chen C.C. and Chui, A. 2010. Positioning strategic competitiveness of green business innovation capabilities using hybrid method, *Expert Systems with Applications*, 38, 1839–1849.

Lundberg, M. and Tell, J. 1998. Networks as a Way of Improving Small and Medium Sized Manufacturing Enterprises, Thesis, Chalmers University of Technology, Gothenburg and Department of Work Organisation, Sweden.

Mariadoss, B.J., Tansuhaj, P.S. and Mouri, N. (2011) Marketing capabilities and innovation-based strategies for environmental sustainability: an exploratory investigation of B2B firms, *Industrial Marketing Management*, 40(8), 1305–1318.

Miles, N. and Tully, J. 2007. Regional development agency policy to tackle economic exclusion? the role of social capital in distressed communities. *Regional Studies*, 41(6), 855–866.

Mitra, J., 2000, Making connections: innovation and collective learning in small business, *Education and Training*, 42(4/5), 228–237.

Murphy, J. 1993. A degree of waste: the economic benefits of educational expansion. *Oxford Review of Education,* 19(1), 9–31.

Novelli, M., Schmitz., B. and Spencer, T., 2006, Networks, clusters and innovation in tourism: a UK experience, *Tourism Management*, 27(6), 1141–1152.

Phillipson, J., Gorton, M. and Laschewski, L. 2006. Local business co-operation and the dilemmas of collective action: rural micro-business networks in the north of England, *Sociologia Ruralis*, 46(1), 40–60.

Reinl, L. and Kelliher, F. 2010. Cooperative micro-firm strategies: leveraging resources through learning networks, *International Journal of Entrepreneurship and Innovation*, 11(2), 141–50.

Reinl, L. and Kelliher, F. 2014. The social dynamics of micro-firm learning in an evolving learning community, *Tourism Management*, 40, 117–215.

Santoro, M.D. and Chakrabarti, A.K. 2002. Firm size and technology centrality in industry-university interactions, *Research Policy*, 31(7), 1163–1180.

Saxena, G. and Ilberry, B. 2008. Integrated rural tourism: a border case study, *Annals of Tourism Research*, 35(1), 233–254.

Simango, C. 2000. Corporate strategy R&D and technology transfer in European pharmaceutical industry: research findings. *European Business Review*, 12(1), 28–33.

Swan, J, Scarbrough, H. and Robertson, M. 2002. The construction of communities of practice in the management of innovation, *Management Learning*, 33(4), 477–496.

Walsh, M., Kelliher, F., Harrington, D. and Lynch, P. 2012. Moving towards a Green Economy: Capitalising on Organisational Innovation Capability to Leverage the Reservoir of Knowledge in Learning Organisations – an Irish perspective, *IFASM Conference Proceedings*, University of Limerick, 25–27 June.

Wang, C.L. and Ahmed, P.K. 2007. Dynamic capabilities: a review and research agenda, *International Journal of Management Reviews*, 9(1), 31–51.

Wenger, E. 1998. *Communities of Practice: Learning, Meaning and Identity*. New York: Cambridge University Press.

Wenger. E.C., McDermott, R. and Snyder, W.M., 2002, *Cultivating Communities of Practice: A Guide to Managing Knowledge*, Boston, MA: Harvard Business School Press.

Wickham, P. 2001. *Strategic Entrepreneurship*, London: Pitman.

Zhu, C. 2012. Student satisfaction, performance, and knowledge construction in online collaborative learning, *Educational Technology and Society*, 15(1), 127–136.

3
Multilevel Engagement: Theory and Practice Integration

Authors Various

Abstract: *Chapter 3 begins by acknowledging that the pursuit of SME green skill enhancement needs to be based on collaborative action brought about through multilevel engagement. It goes on to document the design and delivery of a web-based collaborative space incorporating a virtual learning environment with an electronic discussion board, bulletin board and regular network updates, which operates alongside live interventions to optimise multilevel stakeholder contributory action and initiate relationships within and between regional SMEs and their communities. A series of GIFT case studies 'brings to life' the resultant interactions and activities which facilitate the exchange of experiences, resulting in regional SME connectivity with providers of specialist knowledge, beyond network, regional and national boundaries.*

Kelliher, Felicity and Leana Reinl. *Green Innovation and Future Technology: Engaging Regional SMEs in the Green Economy.* Basingstoke: Palgrave Macmillan, 2015.
DOI: 10.1057/9781137479822.0008.

In the GIFT project, once an interdisciplinary collaborative ethos was established as the appropriate course of action (Figure 2.1, Chapter 2), the team sought to create the conditions which would allow effective green innovativeness to be nurtured in and among SMEs resident in a number of regions, some of which were separated by nation. Of relevance is the process through which these conditions are sought and ultimately how collaborative action is built. The interactive relationships within which regional SMEs innovate (Aylward, 2012; Reinl and Kelliher, 2010), allow SMEs to form green innovativeness in cooperation as shared knowledge is used to improve effectiveness, efficiency and innovation (Lesser and Everest, 2001, p. 46). Therefore, the GIFT team recognised that the pursuit of green skill enhancement needed to be based on collaborative action brought about through multilevel engagement. As such, the notion of contributory action and resultant reciprocity and their link to cooperation and collaboration (Aylward, 2012; Koch, Kautonen and Grünhagen, 2006; Miles and Tully, 2007) are also considered.

As noted in Chapter 2, multilevel engagement facilitates the exchange of experiences by connecting regional SMEs with providers of specialist knowledge, who may or may not be resident in the SME's region. When considering multilevel engagement in this way, the interdisciplinary nature of the GIFT programme sought to provide insights into the potential for theory to contribute to practice and vice versa (Kelliher, Harrington and Galavan, 2010). In theory, regional stakeholders such as government agencies, HEIs, indigenous businesses, economic support groups and regional development agencies are each considered pivotal to successful and sustainable regional development (Döring and Schnellenbach, 2006; Drda-Kühn and Wiegand, 2010; Kelliher et al., 2014). These stakeholders ideally facilitate the development of innovativeness skills and in doing so enhance SME awareness, ability and action to meet the goals of sustainable regional development. Ideally, theory-practice integration should be at local, national and international level, as this objective hinges on an integrated regional stakeholder model (Döring and Schnellenbach, 2006). Therefore, the GIFT team needed to create an environment that would facilitate interactions with a variety of cross-disciplinary experts, support agencies, academics and SMEs, thereby creating shared knowledge by continually moving ideas in and out of the SMEs and their regional network.

As the inherent green skill requirements were likely to cross the boundaries of region/country (Trippl, 2010), natural-physical science and business-technology management, cross-discipline integration was seen as a priority in the pursuit of SME green skill enhancement (Brooks and Ryan, 2008; Nastase, Popescu and Boghean, 2009). It was pertinent to begin by bringing interdisciplinary teams together to facilitate deep conversation between these disciplines and consider the dynamics involved in collective up-skilling – a process involving rural-regional businesses, social enterprises, and public sector and educational institutions with the ultimate purpose of engaging an international community of practice (Wenger, 1998).

Chapter 3 presents a multiperspective lens through cumulative case findings. In doing so, it offers a cross-border, multidisciplinary viewpoint on how to design an integrated approach to regional SME green skill enhancement. The exhibited cases form the basis of a funded cross-border initiative whose foundation is committed to promoting environmental benefit and social progress in a sound regional-rural green economy.

3.1 Collaborative spaces to promote cross-border SME/stakeholder engagement and knowledge exchange

The GIFT programme design took into account the GBICs literature (Lin, Tseng, Chen and Chui, 2010; Walsh et al., 2012), which argues that organisations must strategically and routinely reorganise their innovation capabilities to harmonise innovation with the external environment and wider society. Reticent of the restricted resource base on which SMEs engage (Kaufmann and Tödtling, 2002), a reality that can act as a barrier to external engagement (Kelliher and Reinl, 2009), the GIFT team recognised that businesses' and HEIs' contributory action could be facilitated through both physical and virtual engagement. These interactions are labelled 'collaborative space' in the literature (Lin et al., 2010; Zhu, 2012), a term adopted to describe the physical and virtual fora implemented in the GIFT programme and framework (Chapter 2, Figure 2.1). This collaborative space also provided a means for cross-region/country engagement, as participants could be located elsewhere and still log into the virtual environment.

The GIFT team, in interaction with technical experts, developed a web-based collaborative space incorporating a virtual learning environment with an electronic discussion board, bulletin board and regular network updates, to optimise multilevel stakeholder contributory action and initiate relationships within and between the rural-regional SMEs and their communities. In this virtual world, 'it is the relationships and interactions among people through which knowledge is primarily generated', and thus the 'community takes on new proportions in this environment and consequently must be nurtured and developed so as to be an effective vehicle for [learning]' (Palloff and Pratt, 1999, p. 15).

More than 30 physical and virtual learning events took place over the duration of the GIFT programme. The delivery of these generic and thematically specialised seminars, workshops and virtual discussions facilitated cross-border and cross-disciplinary knowledge exchange and transfer (Cope, 2003; Gulati, 1995; Toiviainen, 2007). These learning events/interactions were observed by the GIFT team as they encompass the process of green skill enhancement and afford insights into community of practice evolution.

As interactions on virtual platforms need to be collaborative to leverage the value of skill and knowledge mutuality (Palloff and Pratt, 1999; Passey, 2014), the discussion board provides for a bimonthly 'expert panel' in which all network members can engage with local, national and international specialists online. Ideally, the members of the expert panel act as knowledge brokers within the virtual learning environment, much like those described in Chapter 2. By including international experts on these panels, knowledge scope and flow are enhanced (Erkuş-Öztürk, 2009) and the potential for spatial blindness is diminished (Brouder and Ericksson, 2013). In this regard, virtual and physical fora provide collaborative space (Zhu, 2012) where geographically distant members, sharing domain-/discipline-specific experience and knowledge requirements can meet and participate in broader knowledge-exchange pools.

As the GIFT programme seeks to promote long-term multilevel engagement that is sustained well beyond the programme duration, the underlying goal is to discover, understand and perhaps improve on existing patterns and order (Ghaye and Lillyman, 1997) in order to promote deeper regional SME engagement with the green economy. A community of practice perspective (Lave and Wenger, 1991) helps in this regard as it is an organising principle surrounding the international green community, which has shared aims and whose members systematically and intentionally explore and consider information from research, from experts and from each other (Shields, 2003).

The following three cases are provided to 'bring to life' the activities which engaged the participating regional SME communities in collaborative spaces to promote cross-border SME/stakeholder interaction (Palloff and Pratt, 1999). A VLE provided the linchpin for embedding the green community of practice (Figure 2.1), wherein business and academic experts, as well as community stakeholders across the disciplines engaged in a virtual space with SME owners. Through cyclical engagement with specific green-based topics, all parties could contemplate business and sector challenges and opportunities:

▸ Case A describes the virtual engagement of one of these groups (representing the sustainable tourism sector) and highlights the critical 'knowledge broker' roles of the VLE facilitator and expert industry guests in this context (Phillipson, Gorton and Laschewski, 2006; Reinl and Kelliher, 2014). The GIFT programme also engaged SMEs in a series of continued professional development (CPD) workshops and networking events in and across each region. These interventions were frequently followed by interactive expert sessions delivered in a VLE with, for example, ecotourism specialists, regional funding and government authorities, biofuel engineers and wind energy experts. These VLE activities sought to build the foundations for cross-border learning relationships and facilitate knowledge exchange activities (Cope, 2003; Gulati, 1999; Toiviainen, 2007).

▸ Case B details an example GIFT training day, which sought to promote sustainable walking tourism in two regions, Ireland and Wales. In documenting this case, the authors offer practical guidance and insights into the development of effective cross-border relationships and multilevel knowledge exchange activities (Brouder and Eriksson, 2013; Halme, 2001).

▸ Case C details the multilevel engagement approach which underpinned the design of GIFT's annual learning showcase. It comprises an international SME study tour, cross-border SME case presentations and an interactive expert plenary panel. The authors detail the evolution of this annual event over the three years of the project's delivery, demonstrating lessons learned along the way.

3.1.1 Case A: Creating a VLE to embed a CoP ethos

Authors: H. Gittins, J. Wall, A. Foley

Overview: The VLEs in BU and WIT (HEIs in each region) provided a significant platform for online engagement between GIFT stakeholders (enterprises, faculty, project executive, statutory and third-sector organisations) and a range of industry and subject specialists. These interactions, facilitated through online chat rooms, forum posts and Adobe Connect sessions, resulted in important collaborations on key sustainable development themes in green tourism, waste management and renewable energy.

Collaborative teaching and learning activities: BU and WIT pivoted their expertise in VLE management (Moodle in WIT, Webex online meetings in BU) to craft customised online platforms for GIFT participants. In WIT, this engagement is in line with the outsourced supply chain delivery approach adopted by the Institute, which reflects the importance of academic institutions' engaging with external experts to bring practice-based solutions to business issues. In BU, the virtual fora provided a space in which key stakeholders and businesses were brought together to share experiences and learn from multilevel stakeholder perspectives. Importantly, this approach sought to provide collaborative learning spaces to facilitate cyclical multilevel cross-border engagement with the aim of nurturing longer-term knowledge exchanges within a CoP framework.

Engagement activity: A series of 'GIFT discussions' included Marketing Walking Tourism Channels: Tools and Pricing; Green Innovation in Tourism; Sustainable Waste Management: Exploring Potential Green Business Opportunities; Sustainable Tourism Business Planning in the Green Economy, and Anaerobic Digestion. Moderated by a member of the GIFT support hub and academic team, the VLE discussions provided accessible engagement for GIFT members and were designed to be cyclical in nature to facilitate a CoP ethos. Typically, 10–15 participants engaged in each online discussion. The virtual learning sets comprised SMEs representing both jurisdictions alongside academics representing the three HEI partners, industry representatives and subject specialists. The learning sets co-produced emergent knowledge, and the academic team ensured that members engaged in a comprehensive learning experience, further supported by cyclical knowledge exchange engagements where appropriate. In the VLE, the questions posed could be responded to in real time, and all present contributed to the generation of ideas, solutions and emergent knowledge. For example, in the walking tourism discussion, specialists from Wales and Ireland provided important insights into live issues such as pricing and market access. Irish participants highlighted initiatives, including the Wrist Band scheme, adopted in Bantry, Co. Cork, in which walkers with the band received local discounts, and participants in Wales discussed progress in innovative areas such as geocaching and printed walking passports.

A series of online discussions run on Webex focused on wood fuel topics and explored business opportunities and the benefits and pitfalls of renewable heat incentive schemes, among other topics. This provided a useful platform to share ideas and best practices across geographically distant borders, and for cross-border network members to learn about the influence of national policy frameworks and resultant financial incentives and the business/market impact in each country. Frequently, these virtual engagements stemmed from knowledge-exchange requests at physical learning events, one such example being a sustainable waste management online question-and-answer session with the Chartered Institute for Waste Management following on from a waste management workshop. As such, SMEs had the opportunity to spend time integrating learning into their own work situations whilst later reconvening to develop and generate further ideas and share their subsequent successes and failures in addressing the challenges of 'real-world' implementation of theory in both jurisdictions.

Outcomes: Important ideas and strategies have developed from the online interactions between entrepreneurs, faculty, subject specialists and policymakers. In the case of the walking tourism discussion, outcomes included a strategy for providing an integrated walking tourism experience through community collaboration, suggestions on the use of digital media to connect with the market, and discussion on the use of college interns to assist with the marketing of walking products. Also, the potential of the 'Walkers Welcome' scheme and digital mapping was highlighted. The VLE activities in the GIFT project reflected the strategic objectives of BU and WIT in generating activity in key areas of technology-enabled outreach, as identified by the Higher Education Funding Council for England (HEFC, 2009), particularly in 'learning resources and environments' in which broad collaboration and effective management of the online learning experience are highlighted.

Testimonial: '*I did appreciate the opportunity to participate in the GIFT Project's online wood fuel discussion with Sam Whatmore from the ConFor Woodfuel Suppliers Group and others involved in the industry on both sides of the Irish Sea and in England. It was particularly valuable to me as a Bangor MBA Environmental Management student and self-employed electricity and gas broker wanting in future to focus on renewable energy rather than fossil fuels. It was good to hear of the real practical business challenges with different grant regimes. All done without having to travel and add to my carbon footprint – fabulous*'. (Entrepreneur and BU MBA student)

Case A exemplifies the translation of VLE knowledge into collaborative action (Palloff and Pratt, 1999; Passey, 2014) – multilevel engagement resulted in a strategy for providing an integrated walking tourism experience and highlighted the potential use of college interns to assist with the marketing of walking products. In turn, the interns gained experience applying the theoretical knowledge generated through their college education, resulting in an appreciation for the challenges and opportunities of practical implementation. It also generated a greater SME understanding of the expertise available within and outside individual regions, thereby expanding the networks' boundary to include international actors (Erkuş-Öztürk, 2009). By connecting regional SMEs with providers of specialist knowledge 'on both sides of the Irish Sea and in England', the VLE facilitated a multilevel exchange of experiences in an international context (Halme, 2001; Zhu, 2012), 'without [participants] having to travel'. This approach involved the active participation of local, national and international actors in a multidisciplinary framework, potentially leading to the establishment of an international 'green' community of practice (Lave and Wenger, 1991; Wenger, 1998).

3.1.2 Case B: Training days: Collaborative engagement to promote sustainable walking tourism

Authors: K. Clayden, E. Young and E. Owens

Overview: Tourism is 'an activity essential to the life of nations because of its direct effects on the social, cultural, educational, and economic sectors of national societies and on their international relations'.[1] The most popular activity in which the overseas visitors to Ireland participated is walking/hiking, with over 775,000 visitors engaging in this pursuit (Fáilte Ireland, 2012) and a market potential of approximately €8 million from overseas walkers alone (Fáilte Ireland, 2013). While this case focuses on the Irish context, this tourism activity is emulated in other rural/regional domains across Europe. In response to the strategic importance of this sector (as identified by GIFT's project partners, including the national Tourism Authority), GIFT, in collaboration with Waterford County Council, organised a conference, attended by over 80 walking tourism business stakeholders, with presentations from Welsh, English and Irish keynote speakers. The aim of the event was to provide Irish SMEs operating in this market the opportunity to gain an appreciation of the walking tourism strategy, hear examples of international best practice walking tourism endeavours and engage in discussions on operational and marketing considerations for developing a sustainable business in this sector, and in doing so, to look at ways of stimulating growth in business within the walking/activity segment of the visitor market.

Engagement activity: Discussions focused on the support requirements of tour operators, walkers and local communities. Ramblers Worldwide Holidays, with over 60 years' experience in walking holiday provision, discussed tour operator expectations from a walking destination perspective and outlined specific market segments that are involved. A case study detailed the walking holiday product and strategy utilised by the organisers of a successful walking festival (Edge of Wales Walking) and highlighted the resultant economic, environmental awareness and social returns gained from those endeavours. Enir Young prompted reflection on the ethos of sustainable business and its impact on the environment now and in the future. Sustainability in both environmental and economic terms was explored, and insights reinforced the value of the multilevel and multidisciplinary stakeholder engagement approach of the GIFT framework. Two interactive discussion sessions provided the opportunity for Irish SMEs in attendance to compare the state of art in a more developed market, that of Wales. This engagement initiated conversations among those in attendance and permitted members to put what they had learned into an Irish context, highlighting the areas of commonality with the UK and the issues which needed to be overcome in order to further develop this sector in Ireland. Members of the support hub noted requests to provide a platform for these discussions to continue after the conference.

Outcomes:

- Differences regarding access law in Ireland and the UK somewhat explain why walking tourism is less developed in Ireland. Wales has many long-distance paths and circular walks where visitors have the freedom to roam independently. The success of Edge of Wales Walking in positioning North Wales as one of the UK's leading walking tourism destinations demonstrates what can be achieved when the walking tourism product is packaged effectively. The Ramblers Worldwide representative was stringent in selecting accommodation, and some destinations are simply too small to attract such operators. This can be an advantage, especially if interested parties collaborate to standardise their offer and work together to attract the more discerning visitor who is attempting to maximise a low volume-high value growth in the sector.
- It is clear from GIFT's engagement in this space that sustainable walking tourism product development requires multilevel stakeholder support, particularly between local authorities and the communities they serve. In this way, realistic growth expectations ensure the capacity of future generations of tourists and host communities to thrive without conflict.
- Notably, several of the Irish attendees exchanged contact details with Welsh guests at the event. Many called for continued collaborative engagement between SMEs, academic and support agencies in the development of the walking tourism product in Ireland and Wales, and beyond. In response the GIFT team planned regular online discussions, training and a cross-border study tour to be held in October 2014 to support potential business development in this sector and nurture further knowledge exchange between these cross-border stakeholders. Furthermore, GIFT partners in Wales are collaborating with Bangor University's SBBS (Business Sense) Responsible Tourism Group via the WISE Network, an ERDF-funded project, providing a basis for sustainable multilevel CoP engagement.

Testimonial: 'The workshop provided good information on business, marketing and gave good examples of best practice with a lot to digest' (Irish B&B owner/manager).

Case B findings articulate that 'sustainable walking tourism product development requires multilevel stakeholder support, particularly between local authorities and the communities they serve' (Brouder and Eriksson, 2013; Halme, 2001), something that was not immediately apparent to participant SMEs. A key finding was that 'several of the Irish attendees exchanged contact details with Welsh guests at the event', thereby breaching each network's boundary to include international actors (Brouder and Eriksson, 2013; Erkuş-Öztürk, 2009). Engagement also highlighted commonalities between the regions, offering evidence that SMEs can readily translate local knowledge (Breschi and Lissoni, 2001) across borders. These findings indicate further potential for cross-border CoP activity (Lave and Wenger, 1991), based on evidence of future (intended) cross-region interactions.

3.1.3 Case C: Annual learning showcase and study tour

Authors: E. Owens and M. McDonald

Case overview: The annual Learning Showcase Event (LSE) and cross-border study tour, held alternately in Wales and Ireland, is an integral part of the GIFT project. The event activities involve a number of Welsh and Irish businesses operating in the green economy and seek to showcase best practice through a series of businesses' presentations followed by a question-and-answer session in each case. The events are typically attended by 80–100 stakeholders and regional business owners. The study tour provides an additional opportunity to build knowledge exchange relationships while providing real-world insights into key sectors of the green economy in both Ireland and Wales.

Collaborative learning activities: An equal number of Welsh and Irish SMEs are selected to represent core GIFT themes and showcase their businesses through a series of interactive sessions facilitated by industry, academic and support hub experts. Business cases highlight successes along with failures and the challenges facing regional SMEs operating in the green economy. In addition, areas of commonality and difference in practice and policy across the regions are debated. By identifying and engaging sustainable business champions in each jurisdiction, the study tour provides participants the opportunity to examine green businesses in situ. Blended together, the LSE and study tour provide an opportunity for Irish and Welsh SMEs operating in the green economy to share best practices, exchange ideas and forge cross-national relationships.

Engagement activity: The first LSE and cross-border study trip was held in Wales. Prior to travelling, the Irish participants had the opportunity to meet and discuss what each party hoped to achieve from the trip, as well as develop an understanding of one another's business. However, the Welsh and Irish representatives did not have an opportunity to meet before or after the LSE. While the event was well attended, with very enthusiastic presentations following a dynamic question-and-answer session, the GIFT team and attendees felt that an opportunity had been missed for greater SME and stakeholder interaction to permit the development of trusting knowledge-exchange relationships across the regions. As a result, time was built into the second LSE to facilitate a pre-LSE networking event at which those present could discover areas of mutual interest and synergy. This resulted in a different dynamic at this LSE, with many members reporting that they were going to 'follow up' with one another after the event.

The businesses involved represented the GIFT themes of sustainable tourism, waste management and green technology, and participants felt that the sectorial range was too broad, with some attendees reporting a feeling that they had little in common with the other businesses present. However, the vast majority of the participants felt that although there was not a direct tie between all of the businesses, the discussions around the event created energy around the green economy sector and provided a network of reinforcement and reassurance to participants regarding their future business development plans.

Outcomes: During the lifetime of the GIFT project, three LSEs have been conducted wherein

- thirty businesses have been showcased to a cross-border audience of 240 SMEs, stakeholders, policymakers and funding agencies;
- three cross-border study trips have been organised; and
- eighteen SMEs and stakeholders participated in the study tours to 15 in situ green business development cases.

Testimonial: 'The event provided a valuable cross national learning experience and networking opportunity for those in attendance and feedback from our network members has been very positive' (GIFT Business Development Officer); 'I've followed up with several contacts I made at the event and I'm hoping that some business will flow from that in the future' (Irish SME – Green Technology).

Case C displays the value of a knowledge facilitator in promoting multilevel engagement (Phillipson et al., 2006; Reinl and Kelliher, 2014). It also exemplifies the potential for cross-border knowledge-exchange activities in pursuit of innovation (Döring and Schnellenbach, 2006). In particular, the study tour allowed for easy academic-practice interaction as and when SMEs requested further support while simultaneously exposing SMEs to their counterparts in other regions/countries. A number of SME owners offered insights from practice in return, facilitating the basis from which trusting relationships could emerge (Miles and Tully, 2007; Reinl and Kelliher, 2010). Notably, this process potentially alleviates the risk that SME owners rely on their own opinion rather than seeking expertise to assess the need for and value of green innovativeness within their business. It also provides a means to release the 'locked in' syndrome discussed in Chapter 2, as 'contacts' made beyond the SME's regional boundary not only mean 'that some business will flow' but also open the potential to look beyond the region's CoP when contemplating sustainable business development.

3.2 Concluding remarks

This chapter sought to document the conditions which would allow effective green innovativeness to be nurtured in and among SMEs resident in a number of regions, some of which were separated by nation. The findings presented in case form offer insight into the process of and value from cross-border collaborative action. Specifically, SMEs gained knowledge and insight from fellow SMEs in their own and other regions, from experts, from academics and from support agencies through engagement with virtual and physical GIFT learning events. The resultant case insights reinforce the view that regional SMEs innovate within sets of interactive relationships. GIFT interactions fostered emergent knowledge sharing among and between learning sets and offered 'good examples of best practice' (Case B) and 'contacts' (Case C) which would have been otherwise unavailable to the SMEs and which they could emulate and possibly improve on in their own business. Whether these interactions will result in improved 'effectiveness, efficiency, and innovation' (Lesser and Everest, 2001, p. 46) remains to be seen.

Note

1 The Manila Declaration on World Tourism, 1980.

References

Aylward, E. 2012. Collaborative Rural Networks (CRNs): An Examination of the Roles and Relationships between Regional Stakeholders, PhD thesis, Waterford Institute of Technology, Ireland.

Breschi, S. and Lissoni, F. 2003. Mobility and social networks: localised knowledge spillovers revisited. Discussion Paper, Università Commerciale, Luigi Bocconi.

Brooks, C. and Ryan, A. 2008. Education for Sustainable Development Interdisciplinary Discussion Series Report. Higher Education Academy. [Internet] Available at: www.he academy.ac.uk/ assets/ York/documents/ourwork/sustainability/interdisc_discuss_series 2008.pdf [Accessed 14 February 2014].

Brouder, P. and Eriksson, R.H. 2013. Staying power: what influences micro-firm survival in tourism? *Tourism Geographies: An International Journal of Tourism Space, Place and Environment*, 15(1), 125–144.

Cope, J. 2003. Entrepreneurial learning and critical reflection: discontinuous events as triggers for 'higher level learning', *Management Learning*, 34(4), 429–450.

Döring, T. and Schnellenbach, J. 2006. What do we know about geographical knowledge spillovers and regional growth? A survey of the literature, *Regional Studies*, 40(3), 375–395.

Drda-Kühn, K. and Wiegand, D. 2010. From culture to cultural economic power: rural regional development in small German communities, *Creative Industries Journal*, 3(1), 89–96.

Erkuş-Öztürk, H. 2009. The role of cluster types and firm size in designing the level of network relations: the experience of the Antalya tourism region, *Tourism Management*, 30, 589–597.

Ghaye, T. and Lillyman, S. 1997. *Learning Journals and Critical Incidents: Reflective Practice for Health Care Professionals*, UK: Quay Books.

Gulati, R. 1995. Does familiarity breed trust? The implications of repeated ties for contractual choice in alliances, *Academy of Management Journal*, 38(1), 85–112.

Halme, M. 2001. Learning for sustainable development in tourism networks, *Business Strategy and the Environment*, 10(2), 100–114.

Kaufmann, A. and Tödtling, F. 2002. How effective is innovation support for SMEs? An analysis of the region of Upper Austria, *Technovation*, 22(3), 147–159.

Kelliher, F. and Reinl, L. 2009. A resource-based view of micro-firm management practice, *Journal of Small Business and Enterprise Development*, 16(3), 521–532.

Kelliher, F., Aylward, E. and Lynch, P. 2014, Exploring rural enterprise: the impact of regional stakeholder engagement on collaborative rural networks. In: C. Henry & G. McElwee (eds) *Exploring Rural Enterprise: New Perspectives on Research, Policy and Practice*, London: Routledge.

Kelliher, F., Harrington, D. and Galavan, R., 2010, Spreading leader knowledge: investigating a participatory mode of knowledge dissemination among management undergraduates, *Irish Journal of Management*, 29(2), 107–125.

Koch, L., Kautonen, T., and Grünhagen, M. 2006. Development of cooperation in new venture support networks: the role of key actors. *Journal of Small Business and Enterprise Development*, 13(1), 62–72.

Lave, J. and Wenger, E. 1991. *Situated Learning: Legitimate Peripheral Participation.* Cambridge: University of Cambridge Press.

Lesser, E. and Everest, K. 2001. Using communities of practice to manage intellectual capital, *Ivey Business Journal*, 65(4), 37–41.

Lin, Y., Tseng, M-L., Chen C-C, Chui, A. 2010. Positioning strategic competitiveness of green business innovation capabilities using hybrid method, *Expert Systems with Applications*, 38, 1839–1849.

Miles, N. and Tully, J. 2007. Regional development agency policy to tackle economic exclusion? The role of social capital in distressed communities. *Regional Studies*, 41(6), 855–866.

Nastase, C., Popescu, M. and Boghean, C. 2009. Promoting entrepreneurship and developing an environment favourable to SMEs, *Annales Universitatis Apulensis Series Oeconomica*, 11(2), 755–760.

Palloff, R. and Pratt, K. 1999. *Building Learning Communities in Cyberspace: Effective Strategies for the Online Classroom.* San Francisco: Jossey-Bass.

Passey, D. 2014. *Inclusive Technology Enhanced Learning.* New York: Routledge.

Phillipson, J., Gorton, M. and Laschewski, L. 2006. Local business co-operation and the dilemmas of collective action: rural micro-business networks in the north of England, *Sociologia Ruralis*, 46(1), 40–60.

Reinl, L. and Kelliher, F. 2010. Cooperative micro-firm strategies: leveraging resources through learning networks, *International Journal of Entrepreneurship and Innovation*, 11(2), 141–150.

Reinl, L. and Kelliher, F. 2014. The social dynamics of micro-firm learning in an evolving learning community, *Tourism Management*, 40, 117–215.

Shields, P.M. 2003. The community of inquiry classical pragmatism and public administration, *Administration & Society*, 35(5), 510–538.

Toiviainen, H. 2007. Inter-organizational learning across levels: an object oriented approach, *Journal of Workplace Learning*, 19(6), 343–358.

Trippl, M. 2010. Developing cross-border regional innovation systems: key factors and challenges, *Journal of Economic and Social Geography*, 101(2), 150–160.

Walsh, M., Kelliher, F., Harrington, D. and Lynch, P. 2012. Moving towards a Green Economy: Capitalising on Organisational Innovation Capability to Leverage the Reservoir of Knowledge in Learning Organisations – an Irish perspective, IFASM conference proceedings, University of Limerick, 25–27 June.

Wenger, E. 1998. *Communities of Practice: Learning, Meaning and Identity.* New York: Cambridge University Press.

Zhu, C. 2012. Student satisfaction, performance, and knowledge construction in online collaborative learning, *Educational Technology and Society*, 15(1), 127–136.

4
Reciprocal Knowledge-Transfer Activities between SMEs and Academia

Authors Various

Abstract: *Chapter 4 contemplates cross-border theory-practice integration as an evolutionary process. It documents the building of multiregional academic-SME knowledge interplays, which can then be expanded to reciprocal innovation knowledge transfer in a cross-border arena. Through a series of case studies, the authors offer a chronology of the three-year GIFT journey, in which the GIFT team mapped knowledge-exchange activities between cross-border multiregional GIFT members. The student-academic-SME interactions presented in the cases exemplify the value of a 'collaborative space', in which each participant has access to different disciplines, insights and, ultimately, knowledge. This collaborative interaction drives green innovativeness and crosses the boundaries of green innovation and future technologies in the natural-physical sciences and business-technology management scholarship.*

Kelliher, Felicity and Leana Reinl. *Green Innovation and Future Technology: Engaging Regional SMEs in the Green Economy.* Basingstoke: Palgrave Macmillan, 2015.
DOI: 10.1057/9781137479822.0009.

As highlighted in the introduction, there is very little research on cross-border innovation fora that deal with rural/regional innovation systems and spaces across Europe (Trippl, 2010). Therefore, pursuit of cross-border theory-practice integration in this arena needs to be an evolutionary process. Successful knowledge exchange between academics and SMEs, resident in multiple locations, is dependent upon the cultural, organisational and management characteristics of the partnership (Dawe, Jucker and Martin, 2005; Kelliher et al., 2010). Thus, the GIFT team goal was to build a series of multiregion academic-SME knowledge interplays, which could then be expanded to reciprocal innovation knowledge transfer in a cross-border arena.

When considering SME knowledge-transfer activities, it is relevant to consider the SME resource criteria and potential barriers to exchange in context. As previously cited, resource constraints often result in SME owners who are focused on immediately applicable performance (Freel, 1999), while competing demands on limited resources can lead the SME to seek only immediately applicable solutions to urgent operational needs of the business (Noel and Latham, 2006). Furthermore, SMEs may fear that the knowledge recipient may use it against them in a competitive environment (Inkpen and Tsang, 2005) or that there may be no benefit offered in return for their own contribution (Dyer and Singh, 1998). In such instances, knowledge exchange may be sacrificed to the detriment of enhanced intellectual resource and, ultimately, SME competitive benefit. This can lead to a self-perpetuating cycle in which limited engagement in knowledge-exchange activities inhibit SME development and may eventually contribute to business failure (Comhar Sustainable Development Council, 2009).

Taking account of the SME literature on knowledge exchange, trust is an enabling factor in accessing resources and facilitating mutual problem-solving, and thus cooperative behaviour that leads to trust is the basis for knowledge transfer and learning within and across network boundaries (Granovetter, 1985; Inkpen and Tsang, 2005; Uzzi, 1997). This ethos promotes a structured approach to small firm knowledge transfer and integration (Mäkinen, 2002; Tell, 2000), in pursuit of green innovation capability enhancement. As SME knowledge is primarily captured through informal discourse and exchange (Reinl and Kelliher, 2010; Tell, 2000), the GIFT team sought to encourage SMEs to engage in informal knowledge-exchange relationships which have the potential to enhance their green innovation capabilities and open resource channels, and/or improve their competitive position (Kearney et al., 2014; Walsh et al., 2012). In the observed cases, open exchange via informal sharing of know-how and reciprocal action ensued once trust had been established among network members (Huggins, 2000; MacGregor, 2004). The resultant knowledge-transfer activities among and between SMEs and academics are documented in this chapter.

The evolutionary nature of knowledge exchange is recognised in this process, as the underlying trust relationship is built through ongoing network connectivity, and SME knowledge is leveraged through 'shared experience' (Kelliher et al., 2014), which allows for contributory and reciprocal action and in turn facilitates mutual understanding. As strong ties offer richer, more detailed and accurate information, and therefore superior informational advantage (Granovetter, 1985; McEvily and Zaheer, 1999; Reagans and McEvily, 2003), the GIFT team sought to encourage strong ties between regional stakeholders. However, in order for members to exchange such valuable information, they must first comprehend that 'cooperation and knowledge sharing can enhance their competitive position' (Inkpen and Tsang, 2005, p. 157) and help alleviate SME resource restrictions through shared resource (Kelliher and Reinl, 2010). Thus, the progressive nature of GIFT programme engagement activities is an important aspect of the journey documented in this book.

SME-academic interactions commenced with the physical and virtual learning interactions detailed in Chapter 3. Subsequently, once trust was established between SME and academic stakeholders over a period of 18 months, the embedded nature of the collaborative practitioner/academic research projects detailed in this chapter facilitated the transfer of more complex, tacit knowledge over time (Hansen, 1999; Uzzi, 1997). It is acknowledged that the relationships documented in this chapter are embedded with trust that encourages GIFT members to share valuable knowledge whilst simultaneously accepting the possibility that this knowledge may be attained by competitors (Dyer and Singh, 1998; Reagans and McEvily, 2003), it is important to note that it took time for this level of trust to be achieved. Thus, the GIFT network provided socially constructed relationships (Lave and Wenger, 1991), which are reliant on embedded knowledge-exchange behaviours, to develop and sustain stakeholder interaction and engagement in the longer term.

Chapter 4 explores the benefits of an integrated stakeholder approach to green innovation and future technology education and knowledge transfer. Through a series of case studies, the authors offer a chronology of the GIFT journey, during which time the GIFT team observed and mapped knowledge-exchange activities between Irish and Welsh network members over a three-year period. The findings suggest that prolonged stakeholder engagement, with appropriate broker involvement (Gulati and Garguilo, 1999; Kelliher and Reinl, 2009), fosters reciprocal innovation and knowledge transfer, ultimately contributing to the development of a cross-border community of practice (Wenger, 1998).

4.1 Postgraduate – regional SME – engagement: Collaborative applied research projects

As highlighted in Chapter 2, collaborative interaction is seen as a priority driver in green innovativeness (Brooks and Ryan, 2008; Nastase, Popescu and Boghean, 2009), as inherent SME skill requirements cross the boundaries of green innovation and future technologies in the natural-physical sciences and business-technology management scholarship. Thus, the GIFT framework seeks to use 'collaborative space' (discussed in section 3.1) to identify cross-boundary collegial projects of mutual benefit to engaged stakeholders. Subsequently, through the fulfilment of small, applied SME research projects (carried out by HEI students under the mentorship of experienced academics), results can be exploited in the short term by the project recipient and in the longer term through knowledge sharing via the GIFT collaborative space. This SME-academic collaborative activity can potentially build trust among all stakeholders (Granovetter, 1985; Inkpen and Tsang, 2005; Uzzi, 1997), ultimately resulting in greater access to deep, tacit knowledge activities and generating stronger regional ties to the green economy.

Based on a practitioner-academic partnership posit, these collaborative projects require a level of partnership above that of traditional engagement (Kelliher et al., 2010; Uzzi, 1997). As such, the GIFT framework and inherent research projects facilitate a focus on a critical SME 'green' problem, thereby facilitating applicable solutions to urgent operational needs of the business (Freel, 1999; Noel and Latham, 2006). Through HEI academia/student engagement, a research plan based on academic methodologies is developed. Strong internal support and commitment help sustain network activity (Human and Provan, 2000) in these projects, and the knowledge broker is a valuable inclusion in context.

Knowledge brokers, in acting as a catalyst for knowledge transfer, can help identify and develop network resources (Kelliher and Reinl, 2009), which 'result from the informational advantages [of] participation in inter-firm networks that channel valuable information' (Gulati and Garguilo, 1999, p. 399). Brokers in the GIFT project, including the programme's academic team and the information/support hub resident in two of the participant HEIs, initially provided safe physical and virtual environments where regional stakeholders could engage with the green community. They then offered a baseline information resource relating not only to the region, but also to cross-regional and European green activities through the programme academics and support hubs. As the programme evolved, these brokers also proved valuable in enabling regional, national and international knowledge transfer (Trippl, 2010), acting as a conduit to informing the wider GIFT community of both the results of and gains made from individual projects.

Guthrie and Warda (2002) have described how leadership is vital for innovation, and top management involvement is a prerequisite to facilitate the knowledge/technology transfer. Thus academic engagement includes practitioner assessment of the proposed solutions in all GIFT projects, and students involved in the mini and maxi projects remain affiliated to an experienced academic tutor for the duration of the research programme. This kind of exposure to actual experiences underpinned by theoretical engagement is gaining credence in the pursuit of management capability development, as it offers a bridge between reflective action and critical theorizing (Starkey and Tempest, 2005; Kelliher et al., 2010). This approach commands an interpersonal aspect to practitioner-academic engagement rather than a formal one, affiliated to SMEs' informal approach to discourse, exchange and knowledge sharing (Huggins, 2000; MacGregor, 2004; Reinl and Kelliher, 2010; Tell, 2000).

A co-education ethos permits all stakeholders, including the academic team and programme participants, to experience the 'learning journey' together via engagement with the intellectual and practical portfolio of interventions developed within the GIFT programme. Thus, this approach blends SME knowledge and academic knowledge, permitting all parties to co-create new contextualised knowledge (Kelliher et al., 2010), thereby enhancing the green innovation skills of both partners. Engaging SME owner/managers and academics in this way also affords an opportunity for the development of new SME and academic capability in reflective practice, skill-theory integration and self-knowledge (Kelliher et al., 2010).

As the inherent green skill requirements are likely to cross the boundaries of region/country (Trippl, 2010), as well as natural-physical science and business-technology management, cross-disciplinary engagement should enhance graduate (and SME) awareness, ability and action in the sustainable development domain (Dawe et al., 2005). This cross-discipline integration should also facilitate the pursuit of SME green skill enhancement (Nastase, Popescu and Boghean, 2009; Brooks and Ryan, 2008). However, as HEIs have traditionally had a low impact on national economies due to inadequate technology transfer methodologies (Murphy, 1993; Starkey and Tempest, 2005), the GIFT programme seeks alternative academic-SME knowledge-transfer activities. Thus, the programme includes undergraduate and postgraduate participation in work-based collaborative projects to facilitate reciprocal knowledge-transfer activities between students (as junior academics), senior academics, policy stakeholders and SME practitioners (Huggins, 2000; MacGregor, 2004).

Peças and Henriques (2006) have described a simple input-output model for collaborative research projects between SMEs and academia, in which the success of the collaborative projects is qualitatively assessed. Indicators of success are based on the level of company involvement, the number of implemented solutions developed and whether SMEs request further collaboration. The initial GIFT focus is on stakeholder interaction and the goal is to leverage this engagement via business-led academe-facilitated skills education and research programmes. Ultimately, the goal is to engage in collaborative business-led green projects to enhance the knowledge-reciprocity cycle (MacGregor, 2004; Huggins, 2000) and the pursuit of trust between regional stakeholders. As such, indicators of success include engagement and evidence of reciprocal knowledge transfer between project participants and throughout the wider GIFT community (Gulati and Garguilo, 1999).

Through cyclical engagement with specific green-based SME projects, reciprocal knowledge transfer could be encouraged between SMEs and academic stakeholders, as documented in the forthcoming sections and exhibited cases.

Sections 4.1.1 and 4.1.2 detail collaborative business-led maxi projects and mini projects, conducted by postgraduate and undergraduate students at the three HEIs involved in the cross-country GIFT project.

These projects were conducted in collaboration with SMEs in the corresponding regions under the mentorship of senior academics.

4.1.1 Business-led maxi projects

Maxi projects are larger-scale research projects focused on business- or industry-led topics within four GIFT themes: Knowledge Economy, Green Tourism, Green Technology and Waste Management. Each project is conducted by a postgraduate researcher, under the tutelage of an academic supervisor/mentor over a six- to eight-week period. Once complete, a substantive business report relating to the findings of the research study is presented to the SME owner/manager, and where appropriate (due to potential commercial sensitivities) the findings are shared with the wider GIFT community. Forty maxi projects are being conducted in years two and three of the GIFT programme.

The following cases provide insights into four exemplary maxi projects:

- HEI 1 project describes a feasibility study on the adoption of eco-trolleys in an SME retail environment in Ireland.
- HEI 2 project examines microbes and methane emissions in the agri-sector.
- HEI 3 project highlights current SME attitudes towards eco-certification in the Welsh context.
- HEI 4 project develops a methodology for an SME green audit.

4.1.1.1 The potential adoption of eco-trolleys by supermarket retailers in Ireland

Author: M. Walsh

Project background: Mann Engineering Ltd. is a family-owned and run subcontract manufacturing and service company based in Wexford, South East Ireland. The company offers complete resolutions for engineering requirements in three core business areas: Precision Engineering, Fabrication/Manual Handling, and Trolley Repairs/Maintenance/Sales. Having heard about the GIFT project through an existing INTERREG 4A project (*Sustainable Learning Networks in Ireland and Wales*), Mann Engineering's Business Development Manager contacted the GIFT support office at WIT to discuss a research project exploring 'green' innovations that would extend Mann Engineering's traditional technological scope. The company was interested in manufacturing more environmentally friendly products and while eco-trolleys 'fit that bill' and were well established across the UK, Europe and internationally, the market for eco-trolleys appeared to be minimal in Ireland, despite supermarket efforts to promote their green credentials and to cut their carbon footprint.

The research objective: Identify the main drivers and barriers that impact supermarket owner/managers' adoption of eco-friendly shopping trolleys.

Methodology: Due to the exploratory nature of this project and respective constraints, four semi-structured interviews were conducted with retail owner/managers in the South East of Ireland. An interview guide was developed based on predefined themes identified in the literature and from initial discussions held with Mann Engineering. The guide was divided into four themes: 'general awareness and knowledge'; 'government and policy interventions'; 'corporate culture and mindset'; and, 'customer engagement and involvement'.

Outcomes: This project facilitated the transfer of knowledge between industry and academia and assisted Mann Engineering to identify potential market opportunities for the supply of eco-trolleys in Ireland. The findings highlight the following main drivers and barriers of adoption:

- A very low level of awareness and understanding currently exists amongst supermarket retailers in Ireland regarding the concept of eco-trolleys; participants called for trials/pilot studies to demonstrate the tangible business benefits of switching to eco-trolleys.
- Timing is crucial in terms of the condition of existing fleets of retail trolleys.
- The cost-effectiveness of investing in a complete overhaul of existing steel trolley fleets was questioned; retailers felt that mixed steel and green trolleys would not be acceptable.
- Finally, participants had no prior knowledge regarding the durability and robustness of plastic trolley alternatives.

Overall, the interview participants expressed a very positive attitude towards the eco-trolley initiative, but emphasised that in order to consider adopting the new trolleys, they would need to be offered appropriate incentives and support from both government and suppliers.

Testimonial: 'Initially the information from the GIFT report would possibly have been seen as negative, mainly due to the lack of knowledge or interest shown by some of the retailers that were interviewed. One year on we would have a different view of the report, as it did highlight that customers needed to be educated in relation to the long term benefits of the eco-trolley both to them and to the environment. There continues to be good response to the trolleys from retailers who have taken the leap of faith and installed eco-trolleys in their stores and it is their stories and experience that we are using to promote the concept with other stores including one large multiple who has taken an interest and will be trialling the eco-trolley later this year. We have also garnered more relevant information from the owners of the concept in Italy and long term feedback from their customers including Carrefour, which is testament to the significant advantages that the eco trolley offers'. (Patrick Coldrick, Business Development Manager, Mann Engineering Ltd.)

This project, while simple in premise, exposed an underlying lack of knowledge relating to expanded green engineering applicable to the SME retail sector, as it was not previously viewed as an immediate priority by the SMEs (Freel, 1999; Noel and Latham, 2006). As retail is among the largest regional business cohorts (Cox et al., 2013; European Commission, 2011), this low knowledge transfer equated to a potential barrier to regional green economy engagement (Uzzi, 1997). Considering the participating SME 'garnered more relevant information' through the project, collaborative interaction is a potential driver in green innovativeness (Brooks and Ryan, 2008; Nastase, Popescu and Boghean, 2009), notwithstanding the perceived need for incentives from regional government agencies.

4.1.1.2 The effect of diet on rumen microbes and methane emissions

Authors: S. Storey, B. Dunne and E. Doyle

Project background: This laboratory-based research project was carried out by Mr Brian Dunne in part-fulfilment of the requirements for successful completion of his studies in the School of Biology and Environmental Science, University College Dublin (UCD) and was supervised by Dr Evelyn Doyle and Dr Sean Storey (GIFT project, UCD). The impetus for the project arose from Department of Agriculture Food and the Marine-funded research showing that supplementation of dairy cow diets with plant oils increased milk yield and decreased emission of methane, a significant greenhouse gas.

Research objective: Methane gas is one of the most potent known greenhouse gases, with a warming potential 23 times greater than that of CO_2. In Ireland, 29.1 per cent of all greenhouse gases is produced by the agriculture sector, with livestock production accounting for approximately 60 per cent of this (Duffy, Hyde, Hanley, Dore, O'Brien, Cotter and Black, 2011). Additionally, methane production represents a loss of up to 8.5 per cent in energy uptake in dairy cattle (Rowntree, Pierce, Buckley, Petrie, Callan, Kenny and Boland, 2010). As a result, Ireland has an obligation under the EU Climate Change Response Bill (2010) to reduce GHG emissions by 2.5 per cent annually to 2020 (Department of Environment, Community and Local Government, 2014). Methane is produced in the rumen during digestion of cellulose by methanogenic microorganisms. Previous research has shown that supplementation of dairy cow diets with plant oils can reduce methane emissions and increase milk yield (Rowntree et al., 2010). The objective of this study was to test the hypothesis that reduced methane emissions were caused by diet-induced shifts in the diversity of methanogen and bacterial assemblages in the rumen.

Methodology: Forty-five Holstein-Friesian cows were allocated to one of three dietary regimes in a randomised block design. All treatments were allocated 17 kg (dry matter) of grazed grass per day per cow plus 4 kg (dry matter) concentrates of stearic acid (Control), soya oil (SO) or linseed oil (LO). The concentrates were offered in equal allocations at morning and afternoon milking. Methane emissions were measured using the SF6 technique (Johnson, Huyler, Westberg, Lamb and Zimmerman, 1994). Rumen contents for microbial community analysis were sampled by means of a cannula and stored at −20°C until use. DNA was extracted from liquid rumen samples using a modified DNA and RNA co-extraction technique (Griffiths, Whitely, O'Donnell and Bailey, 2000) and purified by washing in 70 per cent (v/v) ice-cold ethanol. 16S ribosomal RNA genes were amplified from methanogens and bacteria using primer sets 8F-1492R and 27F-1492R, respectively. Terminal Restriction Fragment Length Polymorphism (TRFLP) (Liu, Marsh, Cheng and Forney, 1997) was used to detect changes in the diversity of the rumen microbial communities in response to diet. Fragment lengths were determined by capillary electrophoresis using an AB3031xl genetic analyser (Applied Biosystems). All analyses were carried out in triplicate. Microbial community profiles were analysed statistically using PRIMER version 6.

Outcomes:

- Methane concentrations from cows varied from 260 g day^{-1} for cows offered stearic acid, 239 g day^{-1} for cows offered soya oil and 221 g day^{-1} for cows offered a linseed oil diet.
- Supplementation of the diet with soya or linseed oil significantly ($p<0.001$) reduced rumen methane emissions, with linseed oil promoting the strongest reduction in emissions.
- Interestingly, supplementation of the diet with either oil did not alter methanogenic communities in the rumen, but bacterial communities were significantly different in cows receiving oil supplemented feed.

These results suggest that the decreased methane emissions observed when cows were fed diets supplemented with linseed and soya oils were due to changes in rumen bacterial diversity, which then impacted on methanogenic activity. This study facilitates a better understanding of dietary effects on ruminal methanogenesis and may give a lead in designing methane mitigation strategies in the livestock industry.

Testimonial: *'The Department remains committed to funding research into the mitigation of greenhouse gases from the agriculture sector. Projects focusing on livestock methane emissions continue to develop and build capacity in this important area of research. The ultimate aim of these investments is to provide the basis for cost effective mitigation solutions that can be easily adopted by the livestock sector'.* (Mr Dale Crammond, Research and Codex Division, Department of Agriculture, Food and the Marine, Ireland)

This project resulted in enhanced technical knowledge relating to the SME agri-sector, a significant economic and social provider in the regional environment. The project offered regional SMEs access to laboratory resources and sought to enhance their technical knowledge (Kearney, Harrington and Kelliher, 2014; Reinl and Kelliher, 2010). The ethos adopted in this project promotes a structured approach to SME knowledge transfer and integration (Mäkinen, 2002; Tell, 2000), in pursuit of innovation capability enhancement. By engaging in co-delivered projects in this domain, not only are SME-academic interactions enhanced but the potential for innovativeness capability development in both partners is increased (Brooks and Ryan, 2008; Nastase, Popescu and Boghean, 2009). The expectation is that this will ultimately improve regional SMEs' competitive position in this sector (Kearney et al., 2014; Walsh et al., 2012), and should result in 'cost effective mitigation solutions that can be easily adopted by the livestock sector'.

4.1.1.3 Attitudes towards eco-certification in the hospitality industry in North West Wales

Authors: H. Gittins, R. Jones, S. Jones and E. Lane

Project background: Tourism is of vital importance to the North Welsh economy, bringing in £1.8bn of annual income and supporting over 37,000 jobs (Tourism Strategy for North Wales 2010–2015; Tourism Partnership North Wales, 2010). Tourism, being the raison d'être for many SMEs, offers many sustainable business development opportunities. Despite this, eco-certifications among these businesses are the exception rather than the rule. The tourism industry has over 100 eco-labels, many of which overlap in methods and scope and are thus challenging to evaluate. The researchers identified this knowledge gap while studying sustainable tourism as part of BU's MBA, coupled with cross-border GIFT network SME discussions which suggested a low take-up of eco-certification schemes in the region in comparison to other areas of the UK. In response to this, BU's MBA student Rebecca Jones investigated visitor and accommodation provider perceptions of certification in North West Wales.

Research objective: The project aimed to provide insights into the perceived benefits and value of eco-certification schemes to both visitors and tourism businesses and to highlight potential barriers to uptake among the tourism business community. Objectives sought to explore

1. customer perception of branding and business strategy in relation to eco-certification;
2. owner/manager ability and willingness to implement eco-certification in their respective businesses.

Methodology: To explore attitudes towards certification, over 200 male and female visitors were surveyed using questionnaires at Betws-y-Coed and Llandudno, popular tourism destinations in North West Wales. In addition, semi-structured interviews were carried out with the proprietors of Bryn Bella, a six-bed guesthouse in Betws-y-Coed, which is participating in the Green Tourism Business Scheme, and the larger Gwesty Carreg Bran Hotel in Llanfairpwllgwyngyll, Anglesey, which is not currently participating in an eco-certification scheme.

Outcomes: The project provided some useful insights into visitor perceptions of eco-certification:

▸ Although respondents welcomed eco-certification, it ranked below location and facilities from a list of destination choice.
▸ Cost was the primary factor for both leisure and business participants, with very few willing to spend more on eco-certified accommodation. This may be reflective of the current economic climate.
▸ Only two per cent of respondents could name a relevant certification.
▸ Seventy per cent stated that they would not be willing to spend more on accommodation certified as eco-friendly.

The findings demonstrate that customers' attitudes do not equate to a drive for sustainability through certification, and also that information relating to eco-certification schemes is neither widespread nor easily available for consumers. The in-depth interviews provided rich insights to add to the above findings. The manager of the larger hotel reported that such schemes were often indicative of better quality and higher standards with the potential to provide a unique selling point *'We need to stand out and do something different – this is a possibility'* (Gwesty Carreg Mon). However, the owner/manager highlighted a need for help and guidance from a third party if they were to consider implementing a scheme, stating that low costs and satisfied customers had to be their priority.

For the smaller operator, they believed 'Certification is just a method of displaying the techniques used within the establishment in order to minimise environmental impacts – savings and cost control just come hand in hand'. The smaller award-winning eco-guest house (certified as the gold standard by the Green Tourism Business Scheme) identified many benefits to accreditation, including real financial savings which result in an 'increase in profits' and an improved ability to 'control costs'. Requirements for mentoring and support at the early stages of the certification process echoed the sentiments of the larger hotel owner/manager.

The case results highlight potential knowledge barriers that exist within a regional community (Dyer and Singh, 1998; Inkpen and Tsang, 2005) and that there appears to be some distance between policy intent (European Environment Agency, 2013; OECD, 2012) and successful implementation of green-based training initiatives. While green policy has increased exponentially (see Chapter 1 for details), successful knowledge exchange between policymakers, academics and SMEs, resident in multiple locations, is dependent upon the cultural, organisational and management characteristics of the partnership (Dawe, Jucker and Martin, 2005; Kelliher et al., 2010). The case findings clearly indicate knowledge-support requirements for small tourism operators in their pursuit of eco-certification and broader sustainable business development efforts, and demonstrate that 'customers' attitudes do not equate to a drive for sustainability through certification, and also that information relating to eco-certification schemes is neither widespread nor easily available for consumers.' This suggests more work is needed to entice green information flow within regions.

4.1.1.4 Developing a green audit for Oceanics Surf School & Marine Education Centre

Author: J. Russell-O'Connor

Project background: Oceanics Surf School and Marine Education Centre is located in the South East of Ireland in the seaside town of Tramore, offering a variety of activities, including surfing lessons, surf camps, coast hacking and environmental education. The main building is a 19th-century single-storey cottage, with a flat-roofed extension and a decking area with two sheds that are used for changing and storage. Oceanics wanted to explore green business strategies and technology solutions that could differentiate them from other activity providers in the area while at the same time providing cost savings. Following initial discussions with SME owner/manager Linda Tuohy, Dr Russell-O'Connor scoped out a project to develop a green audit which would enable Oceanics to operate more effectively in the green economy.

The research objective: A number of standards and energy management systems, including energy audits, have been developed over the last few years that are industrial based, business focused, and that involve onerous levels of paperwork and often require the employment of an energy consultant. The majority of SMEs in the ecotourism sector operate as sole traders, and it is time consuming and expensive for such standards or audits to be conducted in these businesses. While there are no legal obligations at present for SMEs to obtain eco-certification, there is an increasing need to reduce energy use, water consumption and waste production for these businesses not only on the basis of cost but also in terms of their environmental profile and subsequent customer perception. A methodology was developed to assess SMEs in the ecotourism sector, including Oceanics, on this basis, and to identify the feasibility of green technology solutions where applicable. In addition the project sought to identify practices that may be harmful to the environment and propose solutions to minimise or abolish these practices, while enhancing ecological habitats by increasing biodiversity where these businesses operate. Specific objectives sought to

- identify energy-, water- and waste-saving measures; and
- develop a methodology for green audits that could be widely used by SMEs in the ecotourism sector in the UK and Ireland.

Methodology: The audit comprised two phases: an initial survey and an operational period survey. Phase one involved a survey of the building and outside areas, including

- general assessment of energy use, types of waste and annual costs of waste disposal;
- assessment of energy ratings of appliances and frequency of use and a plan for management/monitoring;
- assessment of building fabric – windows, doors, orientation, infiltration and filtration locations, space and water heating;
- assessment of water – number of appliances and locations, age of pipework and possible leaks;
- assessment of lighting – lux meter readings of lit areas and usage requirements.

Recommendations: A number of short-term and longer-term recommendations were made as a result of the audit, including

- compact recyclables more thoroughly and encourage visitors to take home waste;
- repair doors, windows, floors, walls and pipes to combat current energy loss;
- replace bulbs with lower-energy bulbs of appropriate wattage, or in high usage areas consider installing tubular fluorescent lamps, and install timers or movement sensors on lights;
- install flow restrictors on taps and shower head and a timer on shower; log shower use and reduce use during warmer weather; use float boosters in toilet cisterns;

- clean wet suits/decking from collected rain water;
- nominate an employee to monitor, record and review energy, waste and water costs; and
- develop an environmental education garden in unused gravel area – construct plant boxes with pond-dipping container to encourage butterflies.

Outcomes: Recommendations implemented in Oceanics included more effective management of waste and use of energy, and the establishment of a garden for environmental education. The results of the audit highlighted that the energy and water usage for Oceanics was surprisingly low – in comparison to the national average for an SME (SEAI, 2014). Although there were no practices that were harmful to the environment, the audit highlighted requirements for energy and water conservation and staff training to increase awareness. While it was not cost effective to utilize some green technologies in Oceanics, the collaboration increased awareness of energy usage, waste and green technology suitability.

A methodology was developed that could be used by other SMEs without the need to employ an energy consultant and that took consideration of future technologies and opportunities to increase biodiversity. While assessment methods for measuring energy, waste and water use would not be refined for other tourism SMEs, the use of future technologies could be further explored within smaller enterprises with higher levels of energy use. Hence the audit would be refined based on energy consumption and the priorities of measurement. In addition, opportunities for increasing biodiversity should be included in the methodology.

Testimonial: '*Whilst we have always been aware of environmental issues, this study has focused our monitoring of energy and waste more specifically and is now a part of the training given to staff. The initiative has re-enforced and added to our environmental education programme*'. (Linda Tuohy, Oceanics)

The specifics of this project offer insight into the potential trajectory of knowledge sharing relating to the green ethos, and specific programmes of work focussed on the development of particular skills relating to the 'monitoring of energy and waste'. The findings point to the evolutionary nature of academic-practice knowledge transfer and the ongoing enhancement of green innovativeness within regional SMEs. As this project offered more detailed and accurate information (Granovetter, 1985; McEvily and Zaheer, 1999), it has the potential to facilitate superior informational advantage to both Oceanics and the wider regional SME community through the GIFT support hub. However, in order for members to exchange such valuable information they must first comprehend that 'cooperation and knowledge sharing can enhance their competitive position' (Inkpen and Tsang, 2005, p. 157), and thus it should be highlighted that Oceanics' owners have a strong environmental ethos, beyond their commercial intent.

4.1.2 Business-led mini projects

Mini projects are HEI student group projects addressing green innovation and future technology research needs or knowledge requirements identified by SMEs in consultation with multidisciplinary academics on the GIFT project. Project topics span the four GIFT themes: Knowledge Economy; Green Tourism; Green Technology; and Waste Management. The "mini" project is conducted over a four- to six-week period, and generally involves groups of four students working on a specific research objective in consultation with the SME owner/manager and under the supervision of senior academics. These group-based projects are undertaken and assessed in partial completion of a number of accredited green-oriented course modules in the Schools of Science and Business at the three GIFT HEI partner institutes in Ireland and Wales. The findings are disseminated to SMEs and industry stakeholders via physical and virtual learning events and on the GIFT website.

4.1.2.1 *Bachelor of Science (BSc) in Small Enterprise Management*
Author: A. Foley

Project background: The BSc in Small Enterprise Management was developed and designed by the School of Business at Waterford Institute of Technology, in collaboration with Fáilte Ireland, the Irish state tourism support agency. The programme (at level seven- ordinary degree) was the first degree in Ireland targeted at owner/managers of regional micro tourism enterprises, with participation across a number of tourism subsectors, including self-catering, bed and breakfasts (B&Bs), hotels, restaurants, activities (including surfing, bass fishing, yachting) and educational and heritage centres. The programme was distinctive in being both a blended programme (with delivery composed of two-thirds online/one-third on campus) and also demonstrating a Problem-based Learning (PBL) design. PBL as a learning pedagogy presumes a student-led learning process, in which the role of the lecturer is to mentor the students to resolve a business 'problem' or dilemma introduced by the lecturer, with supporting resources and readings allocated by the lecturer in advance. This approach was particularly appropriate for the 23 participants on the BSc, who as busy tourism entrepreneurs were used to adopting a lead in establishing critical business solutions.

Engagement activity: The BSc degree sought to engage the participants with a range of academic and practitioner resources in order to enhance this executive ability. The students also became members of the GIFT network, given the obvious synergy between the environmental remit of GIFT and the myriad of 'green' issues affecting tourism entrepreneurs. These include a desire to utilise sustainable energy sources, practice recycling and effective waste management, and have an allied facility to attract the 'green' tourist, using appropriate market segmentation and communication strategies. The class took a module in 'Regulation and Sustainability' in the second year (with the author as co-lecturer), and a key assignment for this module required distinct groups of students to develop thought pieces (essays of maximum 2,000 words) based on sustainability themes.

Outcomes: This exercise provided a channel for tourism entrepreneurs to offer a précis of current thinking and practice in key sustainable management areas, ending with recommendations for the tourism industry. Importantly, these essays were informed by academic and industry thinking, and then filtered through the direct experience of the participants (for example, Creevy Experience, self-catering cottages of character in Donegal, which has been awarded the EU Flower Eco Label featured as a case example). Therefore, the thought pieces provided a communications nexus on sustainable management between academia, industry, policymaking and tourism practitioners on six sustainable topics:

1 renewable technologies: energy sources (including planning legislation);
2 renewable technologies: waste management (including environmental legislation);
3 green operations and legal implications (tort, contract, employment);
4 green communication/markets (data protection/advertising legislation and self-regulation);

5 green branding and certification; and
6 green business planning (including legal issues relating to sustainability).

Each group presented its specific topic to a review panel, including one of the GIFT executive – and also an industry consultant with GIFT:

▶ The groups posted their thought pieces to the VLE (Moodle) class forum, and the participants found the opportunity to read each of the essays most useful.

▶ The learning interventions were generally designed to enhance the PBL nature of the programme – facilitating the student in taking control of the authorship of the project and guiding the participant to resources which would inform this process.

▶ A subsequent module assessment required the development of individual green action plans informed by thought pieces.

▶ Relevant, up-to-date essays on critical green topics were made available to the wider GIFT learning community.

Testimonial: '*I recently completed the BSc in Small Enterprise Management*... [through] *which I gained a thorough knowledge of business development along with green energy issues... and awareness in sustainability issues, waste management and energy efficiency... We have now set up a "Green Management Team" to identify and implement green initiatives...* [which showed us] *ways of reducing energy loss... energy consumption and energy costs. We are now looking at revamping to further conserve energy output and reduce operating costs*'. (Mags O'Sullivan, Center Manager Aqua Dome)

The PBL ethos pursued in the BSc programme described in this case equates to a structured approach to small firm knowledge transfer and integration (Mäkinen, 2002; Tell, 2000), in pursuit of innovation capability enhancement. By providing a basis for theory-enabled problem-solving within the SME, the HEI offers mutual engagement with the subject matter, and in doing so, provides a means for reciprocal innovation knowledge transfer in a cross-border arena (MacGregor, 2004; Huggins, 2000). The resultant knowledge-transfer activities among and between SMEs and academics have provided a 'channel for tourism entrepreneurs to offer a précis of current thinking and practice in key sustainable management areas', thereby enhancing SME innovation capabilities (Kearney et al., 2014; Walsh et al., 2012).

4.1.2.2 MBA student pitch to SME owners
Author: H. Gittins

Project background: In order to provide Environmental Management (Master's of Business Administration; MBA) students on the Green Technology module the opportunity to assimilate knowledge gained on the MBA course in a real-world setting, students were asked to select and assess energy efficiency or renewable energy options for SMEs and present recommendations in a short written report and through a five-minute presentation to participating owner/managers. In addition to meeting module learning objectives, this approach sought to develop critical employability skills (including consultancy, presentation and report writing skills) for the student's intended future careers in environmental management.

A range of businesses from the GIFT network received academic and topic specialist support on a self-defined business problem or issue. This collaboration offered participating businesses time out from their day-to-day business operations and an opportunity to access fresh ideas from motivated Master's students, guided by senior academics and project specialists. Collaborative exchange relationships developed between SMEs and BU faculty as a result, and offered students real-world experience in preparation for their future careers as environmental management specialists. In the main, participating businesses engaged 0–50 employees and operated in diverse industry sectors. This exercise provided students an opportunity to carry out sector-specific consultancy projects in an area relevant to their own interests and experience, whilst also learning about issues pertinent to other sectors from peer presentations.

Engagement activity: Some excellent examples of benchmarking environmental performance emerged from the projects. The students presented bespoke green technology solutions which offered carbon, resource and financial savings to SMEs, and also outlined strategies to increase competitive advantage through the implementation of a range of sustainability measures. For example, one small hairdressing business received a detailed account of monetary savings that could be achieved by using washable towels rather than the existing disposable version. As recommendations were offered within the parameters of existing company budgets, students developed an appreciation of cross-functional business processes. Participating companies benefitted from the student presentations. Simple draft-proofing measures and low-energy lighting solutions were outlined to an office supplies owner/manager, and staff behaviour-change techniques for electrical equipment use were detailed for a whole-food store business. The feasibility of heat pump and wind turbine installation to meet some of the energy demands of a local engineering firm were explored in another project. A range of options for income generation through renewable technologies was also assessed in the projects. This approach permitted business owners to discuss and evaluate available renewable energy options with one another and with BU faculty.

Outcomes:

- Each company received a summary report recommending tailored solutions and outlining the associated costs and business benefits, in addition to sources of further support for sustainable development.
- Master's students benefited from defining the parameters of projects with real companies and gained valuable experience by presenting their solutions in a persuasive and relevant manner to the participating businesses, many of which had not engaged in and were not familiar with a sustainable development agenda and the associated academic-/discipline-specific language that accompanies it.

Testimonial: '*I enjoyed presenting my findings to the class. I also enjoyed hearing my classmates' presentations as I felt I learnt from them as all businesses were different*' (MBA student). '*The most important thing that I learnt about the subject through doing this assessment was how to apply the technology to businesses*'. (MBA student)

This case highlights the importance of knowing 'how to apply the technology to businesses' and the value of leveraging SME knowledge through 'shared experience' (Kelliher et al., 2014). This permitted students to co-create new contextualised knowledge (Kelliher et al., 2010) with SMEs and with one another by entering the world of the SME and the resource base that determines its development. Underpinning this activity with theoretical engagement resulted in the development of students' managerial capabilities, including persuasion, and negotiation and evaluation skills. As such, this approach bridges reflective action and critical theorising (Kelliher et al., 2010; Starkey and Tempest, 2005).

4.1.3 Case D: Training days: Collaborative engagement to promote wood fuel quality

Authors: T. Kent, M. Kennealy, E. Owens and J. Walmsley

Case overview: The GIFT project team actively sought to collaborate with other national and European projects to identify synergies within areas of mutual interest and ultimately to disseminate best practice outputs in addition to extending the networks of all funded projects.

Following this strategic objective, GIFT in collaboration with the Forest Energy Research Programme (funded by the Department of Agriculture, Fisheries and Food) and Bio-En-Area (funded under the INTERREG 4C programme) co-hosted a research-led, best practice training seminar titled 'Wood Fuel Quality Management'. The seminar was primarily aimed at wood fuel producers and wood fuel stakeholders.

Collaborative teaching and learning activities: The objective of this training day was to engage SMEs, academics, postgraduate students and other sector stakeholders at a regional and international level, to exchange experiences on best practice wood fuel quality management and to collaborate for the development of a sustainable wood fuel sector. The workshop comprised three components. Firstly, a series of expert stakeholder presentations described wood fuel quality management systems in the UK, Italy and Ireland, and also outlined the economic value of the respective wood energy sectors. Following this, research conducted by the Forest Energy programme in the South East region of Ireland with regard to managing wood fuel quality along the supply chain was presented. The final session held in WIT's Carriganore campus lab facility provided a series of demonstrations testing the physical parameters of wood fuel with 'hands-on' opportunities for participants to test wood product quality under the guidance of academic experts.

Engagement activity: The proceedings provided a collective platform to enable participants to engage in dialogue with peers and colleagues regarding wood fuel quality standards. They revealed a number of practical steps required within the fuel production supply chain in order to achieve required European standards of production and operation. Good practice examples from Italy and the UK were reviewed, and a facilitated question-and-answer session provided a comparison with the Irish scenario, ultimately resulting in the identification of steps required to improve the Irish system. The Forest Energy programme presentations also allowed postgraduate students to discuss their research findings with the SMEs present, and practical insights into the applicability and practicality of the research findings in a business setting emerged. During the hands-on practical session, SMEs gained an understanding of what is involved in introducing quality assurance testing in their operation and the associated costs.

Outcomes:

▸ A key lesson emerging from the day's proceedings was that the cost of quality implementation is far outweighed by the cost of non-implementation.
▸ The practical session reinforced the value of a quality system in fuel production across the entire supply chain and the collaborative relationships essential for its success.
▸ This learning event facilitated genuine discussion between SMEs, academics and industry experts working in this sector, many of whom had not collaborated beforehand.
▸ Valuable introductions were made between postgraduate students and wood fuel stakeholders, with one attendee scouting for potential graduate recruits.
▸ Academic researchers presented their work directly to end users and gained valuable insights in relation to the practicality of solutions in working business scenarios.
▸ Notably, network participants from the Bio-En-Area project (which was coming to an end) connected with the GIFT network, permitting continued engagement for all concerned and broadening the potential impact of both projects in tandem.

Testimonial: '*A very good review of wood fuel policy and on-going work. I made new contacts and gained some knowledge on Wood Fuel*' (Irish SME). '*The event comprised an ideal blend of policy, academic and business perspectives. The afternoon session really highlighted the importance of the various characteristics of wood fuel that determine its quality as an energy source. Valuable discussions between Irish and Welsh academic partners led to two Bangor students joining the Forest Energy programme for a green technology research project. This collaboration inspired the two Bangor students to go on to excel in their final year of study and gain impressive graduate roles*'. (Dr James Walmsley, Bangor University)

The co-education approach detailed above illustrates the 'learning journey' of SMEs and academics via their engagement with the blended intellectual and practical portfolio of GIFT interventions, permitting all parties to enhance green innovation skills along the way. While SMEs gained access to seminar rooms, expert panels and laboratory resources, enhancing their technical knowledge and managerial capabilities, academics also entered the SMEs' world (for example, in Oceanics, Mann Engineering and the MBA company participants) and co-developed contextualized knowledge (Kelliher et al., 2010). This approach attempts to close the distance that currently exists between policy intent (European Environment Agency, 2013; OECD, 2012) and the successful implementation of green-based initiatives by creating neutral learning spaces in which both SME and academic experience are valued and optimised. The problem-based learning strategies discussed in 4.1.2.1 further support the resolution of actual business problems with theoretical engagement in pursuit of management competency development, as BSc programme owner/managers developed autonomous reflective action and critical theorizing skills (Starkey and Tempest, 2005; Kelliher et al., 2010), potentially resulting in sustainable SME knowledge exchange flows.

4.2 Developing interdisciplinary curricula in sustainable development education

Interdisciplinary contributions to sustainable development (SD) education were a logical step in pursuit of well-rounded 'green' graduates and professionals. SD is enshrined in laws governing planning and development, construction, forestry management, energy, waste, water and so forth, as a central organising principal in both regions. In addition, current efforts are dually aimed at documenting existing SD provision, as well as facilitating the development of new SD initiatives. There was also significant student demand, as exemplified in the NUS (2012) survey results (which involved 90,000 students at 20 universities in the UK), which revealed that nearly two-thirds of students want to see more about SD in their course curriculum, a finding reinforced by developments internationally (UNESCO, 1998). Nevertheless, significant challenges persist in providing access to students with discipline-prescribed timetables and in demonstrating interdisciplinary study which encourages thinking across traditional disciplines (Brooks and Ryan, 2008; Dawe et al., 2005; Nastase et al., 2009).

Presenting a multiperspective lens offers a cross-border, multidisciplinary perspective on how to implement an integrated approach to green skill enhancement. By doing so, the precedent was set to further embed SD in partnered HEI curricula in the GIFT initiative, in which the lead academic partner has committed to 'embed sustainability and awareness of environmental issues in our curricula across the university' (Bangor University Environmental Policy,[1] 2008). Therefore, one of the core GIFT goals was to expand opportunities for students to access interdisciplinary teaching (Colley, 2009) and international insights. The challenge was to develop transformative teaching techniques that could facilitate this multidisciplinary requirement (Dobson, Hedderman and D'Cruz, 2010; Dobson and Tomkinson, 2012; McEwen, Strachan and Lynch, 2011). This journey is portrayed through a series of case studies:

▶ Case E describes the process of interdisciplinary curricula development in the knowledge-based bio-economy domain.
▶ Case F details physical and virtual learning spaces created for cross-border postgraduate knowledge exchange, drawing from the development of the green technology module. This module was co-developed by the cross-border academic team, which encompassed international student interaction both online and during the international postgraduate study tour.

In each case, the goal was to embed green technology skills in graduates who could work in the next generation of green-enhanced start-up enterprises and regional SMEs. This collaborative interaction acted as a driver in green innovativeness (Brooks and Ryan, 2008; Nastase, Popescu and Boghean, 2009), as inherent SME skill requirements cross the boundaries of green innovation and future technologies in the natural-physical sciences and business-technology management scholarship.

4.2.1   Case E: Greening the MBA

Author: G. Griffiths and M. McDonald

Case overview: The role of the environmental manager in businesses today is in a process of transition from one that traditionally broadly encompassed health, safety and environment, to one that is increasingly complex and strategic. Corporate social and environmental responsibilities are becoming prominent in consumer demand and a prerequisite for tender eligibility. This necessitates a more sophisticated environmental manager who can interpret legislative and audit requirements, and deliver them in a manner conducive to continued economic development, whilst recognising the market trends. This was the starting point for the development of the MBA Environmental Management programme at BU.

Collaborative teaching and learning activities: The aim of this programme was to develop skills in the delivery of economic activities related to environment, green technology and sustainability, by including highly topical case studies from across these sectors reflecting changing strategies and alternative approaches. The course was designed to be suitable for graduates in a wide range of disciplines, including Environment, Engineering, Finance, Social Sciences and other subjects. This caused challenges in developing a learning strategy that essentially assumed little knowledge in some cognate areas, and potentially extensive knowledge and experience in others. However, no group was noticeably disadvantaged, and the students rapidly formed a community of learning with high levels of knowledge exchange.

Engagement activity: The MBA programme was advertised broadly in a range of business, environmental and postgraduate journals and websites. The first intake fell into two main categories. One group had an academic background in Environmental Science or related areas (some being recent BU graduates), whilst the second group consisted of participants coming from the business community, whose roles have required them to develop their environmental knowledge and practices or who wished to green their businesses. The first group were from the outset 'critical friends' whose constructive feedback aided a rapid iteration and development of the programme. For these students, the MBA provided a fantastic opportunity to carry on and expand the environmental knowledge gained in their undergraduate degree(s), putting knowledge into a more realistic context, but also teaching about how businesses function on both a human and a financial side.

There were several activities that were completely new to these students: the development of a business plan, preparation of a green 'pitch' to a company CEO, and the dissertation research, which were all conducted in collaboration with local SMEs. These were diverse, and included topics such as the Circular Economy, employee readiness for change following the decommissioning of Trawsfynydd Nuclear Power Station, the development of a unified energy model, behaviour change and energy efficiency, sustainable Human Resources strategy, Biomass energy generation, sustainability in leading UK supermarkets, the environmental impact of mining, sustainable construction in Wales, the modelling of renewable and low-carbon best-fit energy schemes, enriching communities with sustainable festivals, a photovoltaic roadmap for Wales, eco-certification in the Hospitality Industry and an investigation into the feasibility of fracking. This interdisciplinary development was very much in the ethos of the University, which committed to 'embed sustainability and awareness of environmental issues in our curricula across the university' as a key principle of our Environmental Policy (2013). The proactive approach of the academics in both schools (Science and Business) demonstrated the feasibility of an interdisciplinary study, which encourages thinking across traditional disciplines as a vital part of education.

Outcomes:

▸ A unique programme was developed which embeds the principles of sustainable development and awareness of environmental issues across the university, and to the wider community.
▸ The interdisciplinary delivery has increased the awareness and availability of sustainability in disciplines traditionally weak in such provision. Business School delivery of Strategic Management, Organisational Behaviour, Human Resources and Finance have typically not considered green issues, environmental impacts, legislation and measuring outcomes in terms of the triple bottom line.
▸ Student exposure is maximised to interdisciplinary approaches to research and study. This is carried to the wider community by the involvement of a plethora of local businesses and industry in the dissertation phase of the degree programme, as well as student visits, company presentations and case studies.

Testimonial: '*Studying the MBA Environmental Management is one of the best decisions I have ever made. It has put many things into perspective and I have learnt so much in a short space of time that I will benefit from in the future. The course is like having the best of both worlds, as you are able to access two knowledge-pools*'. (Rebecca Jones, MBA Environmental Management 2012–13)

Case E demonstrates the value of interdisciplinary knowledge exchange in promoting thinking across traditional disciplines and beyond existing 'knowledge pools', thereby offering 'the best of both worlds'. This approach can also release 'locked-in' syndrome (as discussed in Chapter 2, this volume) not only within regions but also within disciplines, as the traditional distance between academics and between academe and practice is challenged and ultimately reduced (Kelliher et al., 2010). Interim multidisciplinary expert engagement, enhanced by cross-border involvement, reassures network users that knowledge flow/generation is optimised, and offers all stakeholders access to and potential for innovation capability enhancement.

4.2.2 Case F: Physical and virtual learning spaces for cross-border student knowledge exchange: Green technology postgraduate study tour

Authors: S. Bond, M. Breen, H. Gittins, R. Griffin, G. Griffiths, D. Harrington, A. Hearne, M. McDonald, E. Owens and J. Walmsley

Case overview: Taking credence in the need to develop interdisciplinary curricula in the knowledge-based bio-economy, the Green Technology Postgraduate Study Tour was jointly delivered by Science and Business academics to a cohort of 61 Science and Business students drawn from nine nationalities and two universities in Ireland and Wales. This cohort was brought together to develop proposals around green technology, based on research-led 'better in class' environmental performance criteria. The goal was to embed green technology skills in graduates who could work in the next generation of green-enhanced start-up enterprises.

Collaborative teaching and learning activities: This study programme involved the development of pedagogies that engaged an international, cross-disciplinary academic team and promoted problem-based active learning approaches in their design and execution. Core to the approach was the concept of bringing diverse student groups together to work on problem-based interventions, which in this case focused on energy and resource management activities in commercial settings. Led by an international academic team, these cross-disciplinary, multinational groups had to navigate the commercial and environmental landscape of a start-up scenario and produce a green-led consulting report that would allow a new venture to generate and brand an initiative to entice funding. Ultimately, 15 business proposals which had the potential to be developed as high-potential start-ups were generated and defended before an expert panel. By our taking this approach to student learning, business skills and knowledge competencies emerged across a distinctive learning platform, as knowledge flowed between science and business students and their tutors in an open learning system. Evidence also highlights the continued interaction between the students' post-study trip and the ongoing interaction among the international interdisciplinary academic team.

Engagement activity: Having been placed in cross-disciplinary teams four weeks prior to tour departure, the multinational groups worked in virtual teams using an eclectic mix of technologies to support their pre-tour communication and workflow. Although the groups had an aggressively short amount of face-to-face time on the tour itself, this pre-established virtual relationship allowed them to finalise their report and prepare for its defence in a single day. Wrapped around this truncated working approach was a series of supports around group development and engagement. For example, the groups met in a social setting the night before the main working event, where team members (who had never physically met) sought each other out, underpinning the value of prior remote engagement. Prior to the expert panel presentations, an inspiring field trip to a proposed £100 million hydropower storage facility in the development phase connected the student projects' proposed intervention with a world-class capital intensive start-up environmental business. A roundtable discussion with the promoter followed in which the students learned that this business, on the threshold of delivery, had emerged from a student competition.

Outcomes: Fifteen consulting reports allowed for a wide range of peer-to-peer interactions with skills shared across disciplines, while the expert panel review afforded students real-life project delivery and negotiation experience. Seventy-six students also honed cross-functional team skills in developing commercially viable environmental businesses plans; the business students mentored their more scientifically minded peers in the commercial and promotional aspects of their projects, and the environmental management students offered insight into their discipline-specific rigour and technical competence requirements to their business colleagues.

Testimonial: *'I wanted to convey my appreciation for allowing me to participate on the GIFT trip to Bangor last week. There was great educational benefit to the trip, I thoroughly enjoyed developing new skills, networking with academics in different disciplines to my own and experiencing the practical application of skills learned to date'.* (MBite Student)

This case demonstrates the value of a phased approach to cross-border/multidisciplinary exposure in pursuit of optimum knowledge exchange among students. Pre-established virtual cross-border exercises resulted in ongoing social and academic interactions among and between student cohorts. These activities permitted geographically distant members to build trust (Granovetter, 1985; Inkpen and Tsang, 2005; Uzzi, 1997), ultimately resulting in greater access to deep, tacit knowledge activities and generating stronger ties to the green economy.

4.3 Concluding remarks

The student-academic-SME interactions detailed in the cases presented in this chapter exemplify the value of a 'collaborative space' (Lin et al., 2010; Zhu, 2012), in which stakeholders were afforded access to 'academics in different disciplines' and students were offered the opportunity to 'experience the practical application of skills learned to date'. This collaborative interaction drove green innovativeness (Brooks and Ryan, 2008; Nastase, Popescu and Boghean, 2009) and crossed the boundaries of green innovation and future technologies in the natural-physical sciences and business-technology management scholarship.

Note

1   Bangor University, Wales is the INTERREG GIFT project lead partner.

References

Brooks, C. and Ryan, A. 2008. *Education for Sustainable Development Interdisciplinary Discussion Series Report. Higher Education Academy.* [Internet] Available at: www.he academy.ac.uk/ assets/York/documents/ourwork/sustainability/interdisc_discuss_series 2008.pdf [Accessed 24 April 2014].

Colley, H. 2009. Education for Sustainable Development and Global Citizenship (ESDGC) – Higher Education Academy Review of a Curriculum Audit in Wales. Wales: Higher Education Academy.

Comhar Sustainable Development Council. 2009. *Towards a Green New Deal for Ireland*. [Internet] Available at: http://files.nesc.ie/ comhar_archive/Comharper cent20Reports/Comhar_25_2009.pdf [Accessed 24 April 2014].

Cox, D., Gagliardi, D., Monfardini, E., Cuvelier, S., Vidal, D., Laibarra, B., Probst L., Schiersch A. and Mattes A. (eds) 2013. A recovery on the horizon? Annual Report on European SMEs 2012/2013. European Commission, October.

Dawe, G., Jucker, R. and Martin, S., 2005. *Sustainable Development in Higher Education: Current Practice and Future Developments Higher Education Academy*. Available at: www.heacademy.ac.uk/assets/York/ documents/ourwork/tla/sustainability/sustdevinHEfinalreport.pdf [Accessed 22 June 2013].

Department of Environment, Community and Local Government. (2014). [Internet]. Available at: http://www.environ.ie/en/Publications/ Environment/ClimateChange/FileDownLoad,25196,en.pdf. Last accessed 24 April 2014

Dobson, A., Hedderman, M., D'Cruz, B. 2010. Opening the conceptual gateway: a multi-faceted approach to transformational learning in a Business School context, *Learning and Teaching in Higher Education*, 5, 114–130.

Dobson, H.E. and Tomkinson, C.B. 2012. Creating sustainable development change agents through problem-based learning: Designing appropriate student PBL projects, *International Journal of Sustainability in Higher Education*, 13(3), 263–278.

Duffy, P., Hyde, B., Hanley, E., Dore, C., O'Brien, P., Cotter, E. and Black, K. 2011. *Ireland national inventory report 2011: Greenhouse gas emissions 1990–2009. Reported to the United Nations Framework Convention on Climate Change*. Johnstown Castle, Wexford, Ireland: Environmental Protection Agency.

Dyer, J. and Singh, H. 1998. The relational view: co-operative strategy and sources of inter-organisational competitive advantage, *The Academy of Management Review*, 23(4), 660–679.

European Commission, 2011. *Observatory of European SMEs*. [Internet] Available at: http://ec.europa.eu/enterprise/policies/sme/facts-figures-analysis/index_en.html [Accessed 13 March 2014].

European Environment Agency. 2013. *Towards a green economy in Europe: EU environmental policy targets and objectives 2010–2050*. Luxembourg: Publications Office of the European Union, EEA Report No. 8/2013.

Freel, M.S. 1999. Where are the skills gaps in innovative small firms? *International Journal of Entrepreneurial Behaviour and Research*, 5(3), 144–154.

Granovetter, M. 1985. Economic action and social structure: the problem of embeddedness, *The American Journal of Sociology*, 91(3), 481–510.

Griffiths, R.I., Whitely, A.S., O'Donnell, A.G. and Bailey, M.J. 2000. Rapid method for coextraction of DNA and RNA from natural environments for analysis of ribosomal DNA-and rRNA-based microbial community composition. *Applied and Environmental Microbiology*, 66, 5488–5491.

Gulati, R., and Garguilo, M. 1999. Where do inter-organisational networks come from? *American Journal of Sociology*, 104(5), 1439–1493.

Guthrie, B. and Warda, J.P. 2002. The road to global best: leadership, innovation and corporate culture, The Conference Board of Canada, *Innovation Challenge*, 1–8 May.

Hansen, M. 1999. The search-transfer problem: the role of weak ties in sharing knowledge across organization subunits, *Administrative Quarterly Science*, 44(1), 82–111.

Huggins, R. 2000. The success and failure of policy-implanted inter-firm network initiatives: motivations, processes and structure, *Entrepreneurship and Regional Development*, 12, 111–135.

Human, K. and Provan, S. 2000. Legitimacy building in the evolution of small-firm multilateral networks: a comparative study of success and demise, *Administrative Science Quarterly*, 45, 327–365.

Inkpen, A. and Tsang, E. 2005. Social capital networks and knowledge transfer, *Academy of Management Review*, 30(1), 146–165.

Johnson, K.A., Huyler, M., Westberg, H., Lamb, B. and Zimmerman, P. 1994. Measurement of methane emissions from ruminant livestock using a SF6 tracer technique, *Environment Science and Technology*, 28, 359–362.

Kearney, A., Harrington, D. and Kelliher, F. 2014. Exploiting managerial capability for innovation in a micro-firm context, *European Journal of Training and Development*, 38, 95–117.

Kelliher, F. and Reinl, L. 2009. A resource-based view of micro-firm management practice, *Journal of Small Business and Enterprise Development*, 16(3), 521–532.

Kelliher, F., Aylward, E. and Lynch, P. 2014. Exploring rural enterprise: the impact of regional stakeholder engagement on collaborative rural networks. C. Henry & G. McElwee (eds) *Exploring Rural Enterprise: New Perspectives on Research, Policy and Practice*, London: Routledge.

Kelliher, F., Harrington, D. and Galavan, R., 2010. Spreading leader knowledge: investigating a participatory mode of knowledge dissemination among management undergraduates, *Irish Journal of Management,* 29(2), 107–125.

Lave, J. and Wenger, E. 1991. *Situated Learning: Legitimate Peripheral Participation.* Cambridge: University of Cambridge Press.

Lin, Y., Tseng, M-L., Chen C-C and Chui, A. 2010. Positioning strategic competitiveness of green business innovation capabilities using hybrid method. *Expert Systems with Applications*, 38, 1839–1849.

Liu, W., Marsh, T., Cheng, H. and Forney, L. 1997. Characterisation of microbial diversity by determining terminal restriction fragment length polymorphisms of genes encoding 16S rRNA, *Microbiology,* 63, 4516–4522.

MacGregor, R. 2004. Factors associated with formal networking in regional small business: some findings from a study of Swedish SMEs, *Journal of Small Business and Enterprise Development,* 11(1), 60–74.

Mäkinen, H. 2002. Intra-firm and inter-firm learning in the context of start-up companies, *International Journal of Entrepreneurship and Innovation*, 3, 35–43.

McEvily, B., Zaheer, A. 1999. Bridging ties: a source of firm heterogeneity in competitive capabilities, *Strategic Management Journal*, 20(12), 1133–1156.

McEwen, L., Strachan, G. and Lynch, K. 2011. 'Shock and awe' or 'reflection and change': stakeholder perceptions of transformative learning in higher education, *Higher Education*, 5, 34–55.

Murphy, J. 1993, A degree of waste: the economic benefits of educational expansion, *Oxford Review of Education,* 19(1), 9–31.

Nastase, C., Popescu, M. and Boghean, C. 2009. Promoting entrepreneurship and developing an environment favourable to SMEs, *Annales Universitatis Apulensis Series Oeconomica*, 11(2), 755–760.

Noel, T. and Latham, G. 2006. The importance of learning goals versus outcome goals for entrepreneurs, *International Journal of Entrepreneurship and Innovation*, 7(4), 213–220.

OECD. 2012. *OECD Environmental Outlook to 2050 – The Consequences of Inaction.* OECD Publishing. ISBN 978-92-64-12216-1 (print), ISBN 978-92-64-12224-6 (PDF).

Peças, P. and Henriques, E. 2006, Best practices of collaboration between university and industrial SMEs, *Benchmarking: An International Journal,* 13(1/2), 54–67.

Reagans, R. and McEvily, B. 2003. Network structure and knowledge transfer: the effects of cohesion and range, *Administrative Science Quarterly,* 48(2), 240–267.

Reinl, L. and Kelliher, F. 2010. Cooperative micro-firm strategies: leveraging resources through learning networks *International Journal of Entrepreneurship and Innovation,* 11(2), 141–150.

Rowntree, J.D., Pierce, K.M., Buckley, F., Petrie, K.J., Callan, J.J., Kenny, D.A. and Boland, T.M. 2010. *Effect of either soya or linseed oil supplementation of grazing dairy cows on milk production and methane emissions.* In: Proceedings of the Annual Conference of the British Society of Animal Science, Belfast, April 2010. pp. 51.

Starkey, K. and Tempest, S. 2005. The future of the business school: knowledge challenges and opportunities, *Human Relations,* 58, 61–83.

Tell, J. 2000. Learning networks – a metaphor for inter organizational development in SMEs, *Enterprise and Innovation Management Studies,* 1, 303–317.

Trippl, M. 2010. Developing cross-border regional innovation systems: key factors and challenges, *Journal of Economic and Social Geography,* 101(2), 150–160.

Uzzi, B. 1997. Social structure and competition in inter-firm networks: The paradox of embeddedness, *Administrative Science Quarterly,* 42(1), 35–67.

Walsh, M., Kelliher, F., Harrington, D. and Lynch, P. 2012. Moving towards a Green Economy: Capitalising on organisational innovation capability to leverage the reservoir of knowledge in learning organisations – an Irish perspective, IFASM conference proceedings, University of Limerick, 25–27 June.

Wenger, E. 1998. *Communities of Practice: Learning, Meaning and Identity.* New York: Cambridge University Press.

Zhu, C. 2012. Student satisfaction, performance, and knowledge construction in online collaborative learning, *Educational Technology and Society,* 15(1), 127–136.

5
Green Stakeholder Engagement: The Learning Journey

Authors Various

Abstract: *Chapter 5 presents a collection of multi-level reflections into the interdisciplinary academic collaboration undertaken by the GIFT team, with contributions from the project's funders, the academic team and GIFT's advisory board. Expert/industry stakeholders and SMEs provide industry-based viewpoints on the GIFT experience and a post-graduate candidate offers a reflection prior to stepping out into the green business environment. These stakeholder insights document the pursuit of a framework of green stakeholder engagement to facilitate the development of effective cross-border relationships capable of generating the knowledge exchange activities required for green skill enhancement. The book concludes with a discussion on the constraints and opportunities inherent in engaging regional SMEs in the green economy before contemplating the next steps in this journey. Finally, a number of recommendations are offered for those seeking to optimise stakeholder contributory action in pursuit of SME innovativeness capability enhancement and sustainable development in regional green economies.*

Kelliher, Felicity and Leana Reinl. *Green Innovation and Future Technology: Engaging Regional SMEs in the Green Economy.* Basingstoke: Palgrave Macmillan, 2015.
DOI: 10.1057/9781137479822.0010.

5.1 Reflections on cross-border interdisciplinary academic-practice collaboration

5.1.1 Academic team

Authors: E. Doyle, G. Griffiths and E. Owens

In 2006, a multidisciplinary team of academics from the School of Biology and Environmental Science (University College Dublin, Ireland), School of the Environment, Natural Resources and Geography (Bangor University, Wales), School of Business (Bangor University, Wales), School of Science (Waterford Institute of Technology, Ireland) and School of Business (Waterford Institute of Technology, Ireland) was assembled. The principle role of the team, as well as directing and supporting all aspects of the project, was to devise, develop and implement a Master's level course focused on green innovation and future technologies in Bangor University, Wales and Waterford Institute of Technology, Ireland.

Shortly after the launch of the project, a meeting was held in Dublin, attended by project staff and project academics. This was the first and only time the whole team got together. This meeting was vital as not all the academics had been involved in the drawing up of the GIFT project. This two-day meeting gave everyone the chance to develop a deeper understanding of the project objectives and the areas of expertise of all the attendees which could be called upon to enable the Master's programme development, CPD development and the business-led projects. Equally important, it allowed the team to identify areas in which outside support would be required to achieve all the project objectives. In order to develop sustainable and comparable Master's-level courses in this area, to be, approved and delivered within the timeframe of the project, it was imperative that an understanding of the course evaluation and approval process for each institute be understood. This initial meeting facilitated this and allowed the team to quickly move on to the development stage.

Although each institution developed its own Master's programme, there is at least 25 per cent commonality across both programmes. This allowed the students and academics the unique opportunity of participating in collaborative multidisciplinary cross-border assignments. For each institution this would have been the first time that this had been achieved, giving the academics a unique opportunity to work with colleagues from different disciplines and institutions, on course delivery and evaluation. The academic collaboration that occurred across all dimensions of the GIFT project allowed a high level of collegiality to develop which produced opportunities for academics, project staff and students, and which extended further than the GIFT project. For example, a BU forestry student carried out his undergraduate project with the Forest Energy Research Group within WIT; WIT GIFT researchers embarked on an exploratory project with the BU Biocomposite Centre; a WIT post-graduate student was granted access to equipment within UCD; and senior academics in WIT and BU collaborated on a further research project which successfully secured Institute of Technology Ireland Research funding in 2013.

While in most institutions, collaboration across the Business and Science disciplines is desired, typically there are not many opportunities for real and meaningful collaboration to occur. The GIFT project has provided the academics with such an opportunity.

5.1.2 Advisory board

The advisory board panel comprised Welsh and Irish representatives from academic institutions, state agencies, small businesses and consultancy firms. Meetings were held quarterly, and due to the cross-border dimension of the GIFT programme, teleconferencing was used to facilitate inclusive participation. Over the three-year programme period the Advisory Panel discussed and directed GIFT themes and objectives in interaction with the wider GIFT project team in Ireland and Wales. Based on feedback from SME and regional stakeholders, sought through event evaluations and surveys, a 360-degree feedback/improvement cycle ensued, supporting the evolution of the GIFT programme and, ultimately, the cross-border community of practice.

Dr Shaun Russell is Director of the **Wales Environment Research Hub** at Bangor University. He is an environmental scientist with degrees from Plymouth and Reading Universities, and he has held lecturing posts at Fort Hare and Rhodes Universities in South Africa, and at the University of Namibia. Dr Russell has worked for the British Antarctic Survey, the University of Kent at Canterbury, and Aberystwyth University. He has been an Advisor with the Prince of Wales Business Leaders Forum, and he regularly works with industry, commerce and the business sectors on environmental management, sustainability and corporate social responsibility issues.

I was honoured to be asked to chair the GIFT Advisory Board at the beginning of the project, as I had seen the benefits that other INTERREG initiatives can bring. I was involved in an earlier cross-border project between the English Channel/ La Manche counties of South East England and the Nord-Pas de Calais region in France. The GIFT Ireland-Wales project has further convinced me of the continuing need for, and deep value of, the ERDF INTERREG scheme.

It has been a pleasure to witness a project that, first of all, steps up to the challenge, and then over-delivers on a quite daunting list of objectives and commitments. The diligence and dedication of the project partners have been impressive, and this has proved infectious and stimulating for the wider audience of beneficiaries and end users associated with the project.

GIFT has achieved its key objective of establishing a cross-border innovation forum in support of a sustainable 'green economy' in Wales and Ireland, particularly through establishing linkages between industry and the region's HEIs. The project has carried out a broad range of technically based economic activities relating to environment, climate change, energy and waste management, and business activities such as sustainable tourism. The lead partners at Bangor University, Waterford Institute of Technology and University College Dublin have brought together and up-skilled businesses, social enterprises and the public sector on both sides of the Irish Sea. As a result, it has established an excellent platform for scaling-up Ireland-Wales co-operation on green business initiatives, far into the future.

It has been gratifying to observe the hard work and important outcomes at the sharp end of project delivery, but also a great experience to work with and learn from colleagues and specialists on the GIFT Advisory Board. This has broken down barriers for me and further opened my eyes to the great value of cross-sectoral as well as interdisciplinary and cross-border work. But probably the best experience for me personally has been the friendships formed as a result of connecting with the enthusiastic and committed project partners, working together to bring the GIFT project to a successful and worthwhile conclusion.

There is one argument in favour of a 'Steering Committee' which observes a degree of distance and independence from the day-to-day work of a project. However, my biggest regret has been the lack of time for me to get more closely involved with many of the GIFT events and activities on the ground in Ireland and Wales. However, I hope to put this right by engaging with the exciting legacy and follow-on work that is already being planned by the project partners, to build upon the solid foundations and far-reaching outputs of the GIFT programme so far.

Dr Yvonne Byrne, Rural Development Consultant. Yvonne Byrne has 20 years' experience in rural development and has been involved in European funding programmes in both England and Ireland. Since establishing her own business in 2010, she has completed work for Local Authorities; Teagasc; WIT; Local Development Agencies; community groups; and small businesses. Areas of work include feasibility studies; design and delivery of training courses; preparation of five-year area-based plans; funding applications; and theme-based strategic plans; and facilitating seminars involving multi-agency partners.

I was involved with the Advisory Panel from the outset, and the aim was to inform and advise on the implementation of the GIFT programme over the three-year period. On reflection of the learning journey, the GIFT programme had a very diverse remit as it had an extensive range of stakeholders with different needs, and it also included a varied array of technically complex environmental issues. Attempting to service all of these requirements within the one programme was always going to be a very challenging undertaking. The green economy encompasses a wide range of issues, including waste, energy, biodiversity, sustainability, climate change and water. Each one of these elements could have been the core focus of this programme alone. However, the aim of this programme was to holistically address these issues together.

Stakeholders in this green sector also have very varying needs and interests. The academic sector's core focus is research and education. In contrast, a small business could either be supplying products in the green technology sector or be attempting to reduce their costs and raise their profile through adopting environmentally friendly innovations. However, due to the transnational dimension of this programme and the large number of experts and business case studies on which to draw, it was often possible to match business problems with relevant academic researchers or comparable businesses that addressed these issues. This was facilitated through the wide range of networking events, workshops, case study tours and online forums that allowed stakeholders to exchange ideas and identify partners from whom they could mutually learn. For this process to be effective, the two regions that were collaborating in this transnational programme need to have mutually compatible green stakeholders to ensure that the exchange of knowledge is beneficial to all parties.

The challenge of the Advisory Panel was to direct the GIFT programme so that it could meet the varying needs of all of the stakeholders. It also needed to identify complementary opportunities between the academic institutions and the business sector, and find mutually beneficial synergies between the two partner countries. On reflection, it was felt that having an Irish Advisory Panel that met regularly, to address the needs of the stakeholders in the south east, would have been more beneficial than the quarterly transnational teleconferencing meetings. An Irish Advisory Panel would have enabled more in-depth discussions to take place and facilitated a greater understanding and analysis of the needs of the green sector in Ireland. The outcomes from these meetings could then be regularly communicated to the Welsh partners to ensure that collaborative opportunities were identified. The discussions from the Irish Advisory Panel would also inform the structure and topics for the transnational showcasing events and case study tours.

Over the programme period, considerable focus was given to the synergies between business challenges in the green economy and solutions that could be explored through research by the academic institutions. This close collaboration between universities and the private sector was highly innovative, and is something that could be developed further in any future European programmes. However, for this process to be effective, it is important to have continuity of members on the Advisory Panel and high attendance rates at meetings. For the GIFT programme to have a long-term outcome in the region, it is recommended that these links between stakeholders both within South East Ireland across the academic, business and agency sectors and also within Wales, where relevant, be maintained.

5.1.3   The support hub

Authors: S. Bond, E. Owens and L. Reinl

As a cross-border Ireland-Wales project, GIFT required a flexible and fairly decentralised approach to both day-to-day managing of the project on the one hand, and also supporting and facilitating a dynamic that enabled useful cross-border collaboration on the other. The 'hub' model, whereby BU and WIT each provided support facilities to their respective regional stakeholders alongside cross-border initiatives sought to fulfil this ethos within the GIFT project.

Therefore, GIFT support hubs at BU and WIT provided a point of contact and an information base for GIFT network stakeholders within and across each region. The GIFT website provided a portal from which all stakeholders could engage with the project's support staff, request and register for CPD programmes and learning events, find out about green economy/thematic/sector developments, request mini- and maxi academic projects, access multiregional/cross-country case studies, and connect with other projects, regional stakeholders, SMEs and academics to develop their skills and businesses in the green economy.

While the project was managed from BU from a funders' perspective, it was in reality co-delivered by a project co-ordinator based in Ireland and a project manager located in Wales, in collaboration with the wider academic team, green experts and support hub executive staff based at offices in BU and WIT. In addition to expertise in the thematic skill areas of green innovation and future technologies, the wider GIFT team had extensive experience of learning and e-learning network design and development, SME innovation capability enhancement and business development, administration, cross-sectoral communications and data protection. This professional secretariat supported the GIFT programme goals and maintained a 360-degree feedback/capability improvement cycle through a multilevel, cross-disciplinary network structure.

As the primary function of the support hub executive is to support reciprocal innovation and knowledge transfer, the entire GIFT team considered the elements of design and delivery that could facilitate these goals. The hubs, in conjunction with academic and expert liaison, had the potential to offer appropriate knowledge broker involvement via physical and virtual interventions. These activities were moderated by a member of the support hub to facilitate cyclical engagement in the thematic areas. As the programme evolved, the hubs provided appropriate responses and potential solutions to SME requests for specific project support. These were initially information requests and subsequently expressions of interest to participate in mini- and maxi projects. This progressive depth of SME engagement reflected the phase of 'knowledge exchange' and levels of trust between respective members of the thematic subgroups of the network and provided an appropriate trajectory for their development in a community-of-practice cycle. The support hub's executive staff initially played an important role in allaying fear of knowledge sharing by creating safe physical and virtual learning environments and maintaining codes of confidentiality in communications. This was not always easy as trust was slow to develop in certain cases, and thus we were ever-mindful of protecting personal/commercially sensitive information when building the trust required to broker sustainable knowledge exchanges between and among SMEs and academia.

While the entire team was concerned about how to translate a complex project plan into functions, activities and support structures, covering two INTERREG regions, five departments and three universities across North West Wales and South East Ireland proved less of a challenge than might have first appeared. Modern telecommunications have shrunk borders, and effective cross-border knowledge exchange and collaboration were extensively supported through a range of digital technologies, including Moodle, Blackboard and Skype, all of which supported cross-border knowledge exchange for geographically distant SMEs interested in learning with and from one another. Indeed it would be impossible to see how such cross-border relationships could be supported without the use of such tools.

The GIFT project had a myriad of up-skilling components, some of which were delivered from a Welsh or Ireland base owing to region-specific knowledge requirements. However, in seeking to optimise a cross-regional knowledge transfer context, the support staff sought to ensure balanced representation from both jurisdictions where appropriate. This resulted in challenges of access and coordination as the hub members' simultaneously agreed SME travel arrangements while negotiating cross-border multidisciplinary expert availability for both physical and virtual learning events. Other interventions required cross-border co-operation and co-development, particularly the study tours and the annual learning and showcase event. The cumulative benefit of these activities was that over time, a cross-border community of practice began to emerge, one which we hope will live on well beyond the GIFT programme's lifespan.

5.1.4 Expert/industry stakeholders

Dr Jonathan Derham, Environmental Protection Agency (EPA) is an NUI sciences graduate with a Master's Degree in Management. Jonathan has worked for over 24 years in the environmental field for public- and private-sector employers in Ireland and the UK, and has vast experience of environmental management in industrial/commercial waste facilities. He is currently employed by the EPA, where he heads up the Climate, Research and Resource Use Programme. Jonathan holds the Chair of the National Waste Prevention Committee, and represents Ireland on a number of EU Resource Efficiency-related working groups. His interests include Sustainable Consumption and Production, Green Economy, Green Innovation, and Circular Economy. He works closely with industrial-sector and trade organisations, and employers' groups in discharge of his role.

My initial role with the cross-border GIFT team was to frame the project's objectives and outputs in context with everything else that was going on nationally in Ireland and more broadly in the EU. Prior to this engagement, I did not fully appreciate that WIT had the capacity and remit for economic development of SMEs in the south-east region, so I made contacts which were new to me in that regard, and I suspect these relationships will have ongoing knowledge-sharing benefits.

Contemplating the initial project launch in Wales, I think it was very successful. The strategic stakeholder support was impressive; BU's Chancellor hosted a diverse but effective selection of speakers. At this early stage of collaboration, it struck me that it was a challenge for those involved to 'pitch' the project, so as a starting point discussions centred on sustainable ideas ongoing in North Wales and how the project could add value in that space. At a subsequent GIFT event hosted in our EPA headquarters, I witnessed a lot of energy in the room, with these disparate businesses chatting with each other socially, hearing ideas and sharing stories about sustainable business projects. This gathering achieved a buzz around thinking of things in a different way. I do not think the outcome of it will necessarily be one business working in collaboration with another, but rather that businesses were re-energised by what they heard and could validate their own decisions so they can say, 'right, I'm not alone, there are other businesses doing what I'm doing'. A central message was that being sustainable (that is, operating in the green economy) can be profitable and hugely personally rewarding.

Reflecting on GIFT's aim, I think that there was a challenge, a challenge to make it real for the day-to-day business of the SMEs involved. Coming together as stakeholders, we can talk about national and EU policy all day long, but what GIFT achieved was stripping away the barriers between SMEs, stakeholders and policymakers, and getting them chatting with each other on a business-to-business level. One of the key learnings for me was that when you design and offer suitable content and a forum for SMEs to share that content, then they will come. There is clearly a hunger for this type of informal learning opportunity amongst SMEs, and the business models and strategies highlighted at GIFT CPD events further support SME development in the green economy.

5.1.5 Participating SMEs

Linda Tuohy founded **Oceanics Surf and Marine Education Centre** in 1997 with her husband, Paul, as Ireland's first dedicated surfing school. Oceanics plays a fundamental role in raising awareness about environmental and ecological issues through their development of numerous educational programmes, including the Green Schools Programme. Linda has been/is actively involved in the GIFT network

and has implemented numerous sustainable development strategies in the family-run business over the last 17 years. Her business is the subject of case 4.1.1.4 in Chapter 4.

Our growth as a business arose from a passion, our love of surfing and the holistic lifestyle which that offered. So green development was something that we did not really recognise when we began the business, but as we developed contacts with people from a sustainable development background, we became more aware of how we fit into the green economy and how we can promote sustainable development as we move our business forward. GIFT is the first network I have been involved with that brings together the business community, academics and support agencies with a specific focus on green business innovation and development, and it has opened my eyes to the green economy.

GIFT provided a brilliant opportunity to link with academics as they tend to think differently from entrepreneurs. The Regulation and Sustainability module I recently completed on the BSc in Small Enterprise Management, and with the GIFT team at WIT, reinforced what we need to be doing as green businesses. As part of the accreditation process for the Green Hospitality Award, we engaged in a green audit, and GIFT provided an online discussion and CPD training in this area, linking us with experts on this topic and others with experience of the various schemes. The accreditation process is very costly and time consuming for a small business, and requires an enormous amount of hard work and commitment. Having the support behind us to push our green credentials, to say, 'this is what we're doing', is hard to do unless you are mixing in the right circles.

The nature of the surfing business necessitates the use of large volumes of water, and one of the biggest sources of waste is water usage, so Paul joined forces with GIFT to develop waste reduction solutions. While we are below the industry average, we now monitor our waste and are more aware of options available to us. We have not yet transferred it to the staff, due to other time commitments, including completion of the BSc, but we will implement them moving forward. We have identified our current market position and are repositioning ourselves, placing greater emphasis on the marine education centre. We are developing the ecology and environmental side of the business internally at the moment, and our identity on the website will reflect our ongoing commitment to the environment.

I think success in this space is about sustainable networking and the building of trust. Reflecting on GIFT's network structure, I think that the cross-border aspect worked. It is important we look outside Ireland and towards Europe. The Welsh are further ahead in certain areas than we are, and maybe in other areas they are not so far ahead. However, I feel that initial contact was not really followed up because there were not many similar small businesses with which to cultivate relationships. What we are doing is not similar to the guy with the wind farm, so for now there is not a tie-in to have ongoing contact, yet when we are brought together with like-minded people, there is an opportunity for us to talk and learn from one another. This was our first step networking for green development, so maybe next time we might develop those initial contacts.

Mark Edwards, Owner of **Bryn**

Bella Guest House, opened the doors of his establishment for the first time in 2003 with his wife, Joan. Bryn Bella Guest House has always worked hard to minimise waste of any kind as both Mark and Joan realised that waste costs money. With that in mind, they set about improving efficiency within the business, which soon set them on a sustainable voyage of discovery. Mark is also a Tourism Ambassador. He visits schools within North Wales, raising awareness of tourism as a career path, and is able to demonstrate at first-hand the advantages of sustainable living to a young audience. Mark has been actively involved in the GIFT network from the beginning and, as a result, improved the day-to-day operation of Bryn Bella Guest House by implementing sustainable policies and deploying new technologies.

We considered a change of pace in our lives and decided that a life in Snowdonia would only be possible as a farmer or as part of the tourism sector. From then on, running a Guest House was the only logical step for us. Our business centres around our passion for the mountains and lakes of Snowdonia, coupled with the quality of life that comes from a rural lifestyle. Cost control was a key factor in business development from day one, and a green way of life came to us as a by-product of our operating style.

In the early days, working with like-minded people and businesses was difficult as there were few around. GIFT was influential as it was the first dedicated networking group of its kind we came across. Through GIFT, we were able to meet up with people we would not have come across in any other way. It exposed us to a wealth of technical knowledge and best practice tips and advice to which we simply would not have had access. Where else can you sit next to a senior manager of a nuclear power plant and discuss the problems we face in reducing the carbon footprint of our respective businesses? However, we would also like to think that we have given ideas to other businesses that have helped them reduce their carbon footprint.

GIFT provides a unique platform enabling us to link with businesses and academics interested in sustainable development. It also enables us to mix with students, who will take the green economy forward in the future. This not only enables us to learn from top experts in the field but helps us influence future thinking based on real-time experience gained first-hand. Nowhere else could we as a micro business gain that level of support and guidance. In that respect GIFT has been invaluable to us.

As part of our waste reduction programme, we regularly monitor utility usage such as water, electricity and oil. Many of the events GIFT has organised centre around this important theme, and the lessons learnt from them have been put in place, which has resulted in significant saving in all three areas. This not only saves us money but considerably reduces our carbon footprint. These savings would have been so much harder to achieve without the regular input from the GIFT project. We have long been identified as a green business, but we are continually pushing the boundaries of the sustainable market. Our current projects involve redesigning our gardens to reflect a more wildlife-friendly perspective, as well as demonstrating just how easy it is to grow your own fruit and vegetables throughout the year. The skills and knowledge gained from both these initiatives and the egg production for our own chickens is regularly passed on to our guests as a way of re-educating and leading by example, demonstrating just how easy a sustainable lifestyle can be with limited resources, confirming our ongoing commitment to the environment.

The fact that the GIFT network takes us outside of Wales has helped immensely as it opens up whole new possibilities to network, share experiences and look at the problems we face from a different perspective. Bringing together many diverse businesses from different countries has highlighted just how different we all are, and yet at the same time how similar our needs are to achieve a sustainable business model. We can think of no other network forum where we can be rubbing shoulders with senior managers of one of the country's largest landowners who is discussing the merits of Marine Heat Source Pumps, and then be speaking to a farmer in the mountains of Snowdonia who is about to install a hydroelectric system. Everybody we have met through GIFT has brought something to the table and left with a volume of knowledge that could not have been gained as easily anywhere else.

5.1.6 Participating students

Rebecca Colley Jones has spent over 17 years working within the environmental sector, specifically in resource efficiency. Rebecca is the WISE Network co-ordinator (an ERDF-funded collaborative project, between Aberystwyth, Bangor and Swansea Universities), developing eco-innovation within SMEs in Wales. Additionally, Rebecca is the director of Ynys Resources Limited, an environmental consultancy and eco-tourism business.

Whilst having extensive experience in the sector and learning many business skills hands on, I felt that a personal weakness was the lack of formalised business training. The MBA therefore provided the perfect opportunity to remedy this. The interdisciplinary approach to the MBA provided an excellent blend of business and environmental modules. The business elements enhanced my understanding of the mechanisms within organisations, and I was able to relate to my own experiences within existing and previous projects I had worked on, identifying where improvements and solutions could have been made. In addition, the access to businesses through the GIFT project allowed an insight into the practical applicability of many of the academic theories and an understanding of business needs in this area. Group work with other MBA students from different disciplines within Bangor and from Ireland added to the experience, providing an international perspective.

As I am already engaged in business development, I was able to apply the new skills I was learning immediately to my work with great effect. Within my employment I am applying these skills to projects such as the development of the West Anglesey Demonstration Zone, working alongside Morlais to engage with the supply chain and stimulate economic development in the area. The work on my dissertation has been an amalgamation of skills learnt on the MBA and prior knowledge gained through my career in resource efficiency. The research I am doing will add to my professional experience, expanding my expertise in a developing field. Through this I have already secured funding through DEFRA/CIWM to support me in getting an international perspective on circular economy and in particular the application of resource-efficient business models within Wales. This has also given me the opportunity to engage and work with a number of organisations and programmes such as the ReBus Life project, to add value to the work that they are doing through additional information. A critical element of this work will be to evaluate both economic and environmental benefits for businesses adopting this approach.

The knowledge gained during my MBA will assist me in securing further funding to support the work and the programmes I coordinate. A greater understanding of management finance helps me greatly when communicating with businesses about the potential for development or improvement of their products, as I can work with them to evaluate return on investment, enabling us collaboratively to make more informed decisions. It can also assist in rejecting or postponing solutions as the climate is not yet right.

I hope to present a paper on the outcomes of my research work at an international conference, something I would not have considered doing prior to pursing the MBA. I will also get the opportunity to present my results to DEFRA and CIWM during a forum for those who have received the Master's award. I am the first student from Bangor University to secure this award. I intend to further develop the research I am doing through my dissertation and work in collaboration with a number of Welsh companies which have already expressed an interest, to implement resource-efficient business models.

The MBA has been pivotal in helping me, both as a professional and an individual, expand my horizons and develop a greater understanding of the importance of the combined application of business and environmental skills in the future development of business. As resource resilience becomes a key concern of business, the skills I have learnt will become increasingly important.

5.2 Where next? Key insights and future directions

The GIFT project sought to encourage a green economy in the INTERREG regions of Ireland and Wales, especially the development of green innovation, green tourism, green technology and waste management enterprise (Welsh Assembly Government, 2006; Building Ireland's Smart Economy: A Framework for Sustainable Economic Renewal, 2008). Core to this objective was the creation of a cross-border multi-disciplinary network environment in which regional SMEs and other stakeholders could generate green innovation knowledge and skills and ultimately enhance green innovativeness in this own businesses. A key strategy was to engage regional SMEs in the implementation of a 'new green deal', to move them and us away from 'fossil fuel-energy production through investment in renewable energy and to promote the green enterprise sector and the creation of 'green-collar' jobs' (Building Ireland's Smart Economy: A Framework for Sustainable Economic Renewal, 2008, p. 7). Realising the full benefits of the new green deal depends on making investments in capacity building and providing the necessary skills, training and education required to take maximum advantage of the sustainable jobs of the future. In response, the GIFT team sought to enable a supportive learning environment across communities of practice in both Ireland and Wales in order to take advantage of the possibilities outlined above. A number of key insights are now drawn from our endeavours, and future directions are outlined to further pursue the cross-border innovation goals of sustainable regional development in a green economy.

Some important learning dimensions have arisen as a consequence of the work undertaken on the GIFT project. In the preface, the academic-practitioner exchange that can be encouraged in projects of this scale was highlighted. Indeed GIFT was developed as an instrument to help establish a cross-border innovation forum to grow a sustainable 'green economy' in Wales/Ireland, responding to the Operational Programme call for 'more and better jobs...through greater linkages between the region's higher education institutions and industry' (High-Level Group on Green Enterprise, 2009). GIFT was centred on that report's recommendations for 'ways that encourage stronger industry-academic links [that] can help meet the skills needs of the sector'. For example, 'Education providers should continue to involve industry in the development of course curricula to ensure relevance of skills to employers' (p. 38). The GIFT Continuing Professional Development programme, learning showcases and Master's programmes have each responded to this call for 'more structured engagement between education providers and industry' with CPD opportunities for professionals to update existing knowledge and acquire new skills which enhance the ethos adopted by the GIFT team.

The GIFT approach blends SME knowledge and academic knowledge, permitting all parties to co-create new contextualised knowledge (Kelliher et al., 2010), thereby enhancing the green innovation skills of both partners. With reference to student engagement, having implemented theoretically sound projects in practical settings, these candidates have bridged reflective action and critical theorising, thereby developing much-lauded 'generic skills' in pursuit of management capability development (Starkey and Tempest, 2005; Kelliher et al., 2010). A real success of this project has been the evident employability of the first cohort of MBA students in Wales and the 'GIFT' engaged MSc Business, Innovation, Technology and Entrepreneurship (WIT) and BSc/ MSc (UCD) students in Ireland. These graduates have been 'up-skilled' to work across a range of sectors, armed with both business and environmental skills, having mixed with businesses and carried out cross-border collaborative projects with the network throughout their respective Master's programmes. Specific success stories include graduate roles as a Sustainability Project Co-ordinator with Change Agents UK, a Waste Project Officer with a County Council, a Learning and Development role at Magnox and an Environmental Business Advisor position and the implementation of extensive green initiatives in SME businesses through owner/manager engagement with the BSc in Small Enterprise Management programme. GIFT has also worked closely with industry to deliver a number of business-led collaborative mini- and maxi projects, some of which are detailed in Chapter 4. Here, the GIFT team in collaboration with post-graduate students and academic supervisors from the Schools of Science and Business in UCD and WIT have conducted research in a number of green business areas. These research projects have helped SMEs cultivate strong links with the academic research community to develop or expand a green aspect of their business, look for a solution to a specific problem or investigate a green business opportunity.

In terms of the future, the 'mobility of researchers between academia and "green enterprises" could be encouraged through mechanisms such as developing doctoral programmes in partnership with industry and provision of entrepreneurship training for researchers' (High-Level Group on Green Enterprise, 2009, p. 38). An inherent element of the GIFT Master's programmes, SME study tours and learning showcases is cross-border interaction between regions (Wales and Ireland). These study trips and learning events, documented in case format in Chapters 3 and 4, were actively supported by the GIFT INTERREG research project. The GIFT programme, as exemplified in these case studies, has shown the value of encouraging research and CPD that crosses academic departments, schools, research institutes and borders/ nationalities. These dynamics encourage broader knowledge transfer, beyond regional boundaries, thereby permitting stakeholder access to new forms of learning in which businesses can be exposed to cutting-edge international ideas in scientific, business and educational contexts. From these new and evolving innovation fora emerge new CPD programmes, academic and applied courses, modules and, most importantly, communities of practice. The learning network participants have created a unique space wherein they can seamlessly cross traditional borders through the use and support of smart technology and expert/academic leadership to successfully collaborate with other businesses from (dis)similar backgrounds, industry experts (public and private sector), academic specialists, future graduates and social enterprises. The fundamental goal of the GIFT programme lies in sharing experiences, transferring knowledge across a broader spectrum and creating new opportunities for green growth and sustainable development within regions throughout Europe.

Echoing the sentiments of participating SMEs, GIFT provided the opportunity to connect like-minded businesses to share experiences and inspire ideas. Championing the work of pioneers through case studies, site visits and learning showcases has promoted knowledge sharing based on real working strategies. These demonstrate sustainable businesses which are fully engaged with current and future environmental challenges and opportunities, and are reaping business benefits as a result of their green activities. Here the project cycle of many European-funded projects (usually three–four years) is problematic given the time it takes to build the foundations of SME networks and develop trust to a sufficient level in order to generate a sustainable knowledge exchange dynamic.

The success of the network as a community of practice was dependent on the optimisation of both multilevel and cross-border input, the coordination of which was challenging in terms of time and resources. The VLE and other virtual communication tools detailed in the book contributed to alleviating some of these barriers, although partner and GIFT member dedication was vital in the delivery and immersion of this ambitious programme. The executive support hub staff, being regularly at the front line of GIFT engagement, captured and purveyed ideas and information requests; moved ideas and knowledge in and out of the network; and connected the relevant multilevel stakeholders in pursuit of new knowledge and innovation capability enhancement. In addition to an understanding of an informal yet structured approach to knowledge exchange, as is pertinent in an SME context, this activity required rigorous communication and multilevel reporting via a 360-degree feedback cycle.

The scale and diversity within and between the thematic GIFT areas were both challenging and invigorating in terms of effective CPD delivery. Engaging local champions proved effective in this regard. These individuals, as in situ brokers, knew first-hand the strengths and challenges within their sectors/regional communities and were effective influencers in these domains. Maintaining cyclical engagement in the various thematic groups, some regional and others cross border, was also difficult, as highlighted in the executive board reflections. Many knowledge-exchange relationships were at different stages of development and required tailored responses and appropriate trajectories of engagement. Importantly, establishing shared meaning among the GIFT network as a whole was crucial to ensure that the relevant themed groups remained 'outward looking', as being connected to a larger community of practice they could maintain access to a broad, strategic and multidisciplinary knowledge base.

The project brought existing partners (government, policymakers, business and society/community) into a new realm of work together as a community of practice. Success is evident in requests from participating stakeholders to sustain ongoing interactions with academia to better meet business, client and environmental concerns. This was evidenced over the life of the project and exemplified through the case studies presented in this text. Indeed many companies have continued to have a working relationship with the GIFT team and with each other, and through these engagements they continue to see the value of the practice-led academic interventions and research that this project has sought to institutionalise. However, extension beyond this is hampered by a range of challenges, some previously discussed alongside others, which are summarised here:

- ▸ While consumers and businesses perceive 'greening' as a good approach to business development, there remains a lack of comprehension as to how it really relates to their lives and business in any substantial way. So while awareness of the need to meet emerging green requirements has increased among regional SMEs, businesses and consumers are still reluctant to make significant changes unless they are (1) immediately applicable and (2) do not have cost implications.

- The resource limitations and owner-led local focus of regional SMEs are especially relevant when looking at green challenges beyond today.
- When the project was established, the Welsh team envisaged many benefits from linking the relatively poor Welsh INTERREG region (EU Convergence status in place) with a national capital (Dublin) and with Waterford, situated in the south-east region, having a strong presence of relatively large, hi-tech companies with global links. Perceptions of the Irish economic situation may have proved a barrier to Welsh partners when looking to the Irish experience and taking full advantage of the larger economy and the innovative new ideas there. Ireland's economic recovery provides valuable lessons in the role of sustainability in economic recovery and growth.
- Fortunately, the environmental science offering of the project was very broad, and included traditional disciplines of environmental science (Bangor), chemistry (WIT) and Biology (UCD). The two business schools represented complementary but different skills sets. However, any development in a complex and uncertain world needs access to all kinds of expertise, and it is not known which expertise can be crucial at which point or, indeed, whether the juxtaposition of these can catalyse new thought, encourage new green innovation and innovativeness, or discourage incorrect actions.
- The absence of the centrality of the GIFT programme to bring wider disciplines (which could include philosophy, arts practice, music and visual arts, history, psychology) into the equation in the future may lead to a sense that the GIFT heartland is restricted to the environmental science and business disciplines.
- The lack of a professional 'consultancy'-based instrument in the mini- and maxi research project product portfolio may have been a barrier to some businesses taking part in collaborative research projects in a deep way. For many companies that are not university linked, the time or perceived risk associated with engaging in student projects could have been a barrier to engagement, although is difficult to measure in light of full uptake of available students. Regardless, the addition of a consultancy instrument could potentially encourage SME engagement with the bright fresh minds of students, although mutual expectations would need to be carefully managed.

Reflecting on learning spaces to promote knowledge exchange and ultimately innovation capability development, we can conclude that knowledge resides where social interactions between the right 'mix' of individuals are nurtured. This is not necessarily within a physically or geographically bounded seminar or conference room, nor is it solely the domain of the SME business setting. The cases presented in this book clearly demonstrate the value of contextualised knowledge as academics step into the world of SMEs and vice versa. As such flows of knowledge are 'inextricably linked to the social relations which develop through shared practice' (Swan, Scarbrough and Robertson, 2002, p. 479). Through our interactions, lab visits, study tours, research projects, site visits, case studies and virtual encounters, we co-create knowledge and we learn from one another.

In summary, a key conclusion from our collaboration is that there are no projects, only partnerships. This lesson extends across the HEI academic partnership and the academic-SME partnerships of the GIFT programme. Of note, the European Universities Association (EUA) developed responsible partnering guidelines,[1] and among the highlighted requirements for effective partnering, trust, understanding and time stand supreme. In successfully closing the knowledge-exchange expanse between the INTERREG regions of Ireland and Wales, the GIFT programme and resultant regional stakeholder framework have also abridged the distance between policy intent (European Environment Agency, 2013; OECD, 2012) and the successful implementation of green-based capability development initiatives. In doing so, this book makes a valuable contribution to both research and practice in cross-border green innovation fora for regional innovation systems and collaborative learning spaces.

Note

1 http://www.eua.be/fileadmin/user_upload/files/Publications/Responsible_Partnering_Guidelines_09.pdf [Accessed 25 March 2014].

References

European Environment Agency. 2013. *Towards a green economy in Europe: EU environmental policy targets and objectives 2010–2050*. Luxembourg: Publications Office of the European Union, EEA Report No. 8/2013.

Government of Ireland. 2008. *Building Ireland's Smart Economy: A Framework for Sustainable Economic Renewal*. ISBN 978-1-4064-2244-3, [Internet]. Available at: http://www.taoiseach.gov.ie/BuildingIrelandsSmartEconomy_1_.pdf.

Kelliher, F., Harrington, D. and Galavan, R., 2010, Spreading leader knowledge: investigating a participatory mode of knowledge dissemination among management undergraduates, *Irish Journal of Management*, 29(2), 107–125.

OECD. 2012. *OECD Environmental Outlook to 2050 – The Consequences of Inaction*. OECD Publishing. ISBN 978-92-64-12216-1 (print), ISBN 978-92-64-12224-6 (PDF).

Report of the High-Level Group on Green Enterprise. 2009. *Developing the Green Economy in Ireland*. Department of Enterprise, Trade and Employment, Forfás. [Internet]. Available at: http://www.forfas.ie/media/deteo91202_green_economy.pdf.

Starkey, K. and Tempest, S. 2005. The future of the business school: knowledge challenges and opportunities, *Human Relations*, 58, 61–83.

Swan, J, Scarbrough, H. and Robertson, M. 2002. The construction of communities of practice in the management of innovation, *Management Learning*, 33(4), 477–496.

Welsh Assembly Government (WAG). 2006. *A Science Policy for Wales – The Welsh Assembly Government's Strategic Vision for Sciences, Engineering and Technology 2006*. ISBN 0 7504 9032 2. [Internet]. Available at: http://www.swansea.ac.uk/media/media,40073,en.pdf.

Index

academia
    academic team of GIFT, 92–3
    Bachelor of Science (BSc) in Small Enterprise Management, 73–5
    greening the MBA, 81–4
    green technology postgraduate study tour, 84–6
    MBA student pitch to SME owners, 76–7
    see also knowledge-transfer activities
advisory board, GIFT team, 93–7

Bachelor of Science (BSc), Small Enterprise Management, 73–5
Bangor University, 50, 79, 81, 86n1, 92, 93, 106
Bond, S., 84, 97
Breen, M., 84
brokers, knowledge exchange, 31–2, 36n3, 60–1
Bryn Bella Guest House, 68, 102
Business-led projects
    Bachelor of Science (BSc) in Small Enterprise Management, 73–5
    dairy cow diets and methane emissions, 65–8
    eco-certification in hospitality industry, 68–70
    eco-trolleys by supermarket retailers, 63–5
    green audit for Oceanics Surf School & Marine Education Centre, 70–3
    MBA student pitch to SME owners, 76–77
    promoting wood fuel quality, 78–80
    see also knowledge-transfer activities
Byrne, Yvonne, 95–7

case studies
    creating virtual learning environment, 56–8
    greening the Master's of Business Administration (MBA), 81–4
    green technology postgraduate study tour, 84–6
    learning showcase and study tour, 51–3
    walking tourism, 48–51
    wood fuel quality, 78–80
circular economy, 82, 99, 105
Clayden, K., 48
climate change, man-made, 13
collaboration, *see* GIFT team
collaborative engagement, sustainable walking tourism, 48–51

collaborative teaching/learning
  annual learning showcase and study tour, 51
  case study creating virtual learning environment (VLE), 46
  greening the Master's of Business Administration (MBA), 82
  wood fuel quality, 78
Comhar
  Green New Deal, 18
  SDC (Sustainable Development Council), 18, 22n2
community of practice (CoP)
  cross-border stakeholder engagement, 34
  virtual learning environment for ethos, 46–8
cows, *see* dairy cow diets
cross-border approach
  framework for stakeholder engagement, 34
  green innovation, 33–35
  see also GIFT team

dairy cow diets, methane emissions, 65–8
Derham, Jonathan, 99–100
Doyle, E., 65, 92
Dunne, B., 65
dynamic capabilities theory, 29

Eco-certification
  hospitality industry, 68–70, 83
  Oceanics Surf School & Marine Education Centre, 70–3
economic growth, European Environment Agency, 13
Edge of Wales Walking, 49, 50
education
  Bachelor of Science (BSc) in Small Enterprise Management, 73–5
  developing interdisciplinary curricula, 80–6
  greening the MBA, 81–4
  green technology postgraduate study tour, 84–6

MBA student pitch to SME owners, 76–7
see also knowledge exchange; knowledge-transfer activities
Edwards, Mark, 102
engagement, *see* GIFT team
engaging regional SME's, 3
  regional innovation system, 3
  systems approach, 3
Environmental Outlook, OECD, 13–14
Environmental Protection Agency (EPA), 99–100
EU Climate Change Response Bill (2010), 66
European Environment Agency, 13, 16, 112
European Universities Association (EUA), 112

Foley, A., 46, 73
Forest Energy Research Programme, 78, 79, 93

GBICs (Green Business Innovation Capabilities), 29, 43
GIFT (green innovation and future technology)
  academic-SME knowledge interplays, 58–60
  annual learning showcase, 45–6, 51–3
  case studies, 45–6, 78–80
  collaborative spaces for knowledge exchange, 43–53
  education, 6
  framework, 31–2, 60–1
  goal, 28, 58
  multilevel engagement, 42–3, 44–5
  programme overview, 7–8
  SME perspective, 28–30, 34–5
  themes, 34–5, 63, 73
  training days, 45, 78–80
  virtual learning engagement, 45, 46–8
  see also GIFT team; knowledge-transfer activities

Index

GIFT team
  academic team, 92–3
  advisory board, 93–7
  challenges for, 111–12
  cross-border academic-practice collaboration, 92–106
  expert/industry stakeholders, 99–100
  future directions for, 106–12
  participating SMEs, 100–5
  participating students, 104–6
  Russell, Shaun, 93–5
  support hub, 97–9
Gittins, H., 46, 68, 76, 84
green audit, Oceanics Surf School & Marine Education Centre, 70–3
green business, Ireland and Wales, 21–2
green community, knowledge exchange, 30–2
green economy, 2
  blueprint for, 2
  changing economic environment, 14–15
  collaborative action, 15, 29defining, 15–17
  engaging regional SMEs in, 3–5
  policy gap, 17–20
  transition in regional SMEs, 20
  Wales and Ireland, 94
green growth, 16
Green Hospitality Award, 101
green innovation
  cross-border stakeholder engagement, 34
  generation and integration of regional, 33–4
  wood fuel quality, 78–80
green innovativeness, 1, 3–7
Green Investment Bank, UK, 8n1
Green New Deal (GND), Comhar, 18
Green Schools Programme, 100
green skills
  exchange of experiences, 42–3
  requirements, 62

green stakeholder engagement, *see* GIFT team
Green Technology
  GIFT theme, 34, 63, 73
  MBA student pitch to small-to-medium enterprise (SME) owners, 76–7
  postgraduate study tour, 84–6
Green Tourism, GIFT theme, 34, 63, 73
Griffin, R., 84
Griffiths, G., 81, 84, 92
Gross Value Added (GVA), Wales, 19

Harrington, D., 84
Hearne, A., 84
HEFC (Higher Education Funding Council for England), 47
HEIs (higher education institutes)
  cross-border approach, 33, 35
  GIFT team, 62, 94
  green economy, 20
  stakeholders, 28, 29, 42
  sustainable development, 6
  virtual learning, 46, 61
hospitality industry
  Bryn Bella Guest House, 68, 102
  eco-certification in, 68–70, 83

IISD (International Institute for Sustainable Development), 19
innovation
  cross-border, and knowledge exchange, 30–2
  see also green innovation
INTERREG-funded initiative
  GIFT (green innovation and future technology), 7, 63, 78, 86n1, 94, 107, 109
  regions, 98, 111, 112
Ireland
  Bachelor of Science (BSc) in Small Enterprise Management, 73–5
  dairy cow diets and methane emissions, 65–8
  eco-trolleys, 63–5

Ireland – *continued*
  green business landscape in, 21–2
  green economy ambitions, 18
  green technology postgraduate study tour, 84
  Learning Showcase Event (LSE), 51–3
  University College Dublin, 65–8
  wood fuel quality, 78
Italy, wood fuel quality, 78

Jones, Rebecca, 68, 83, 104–6
Jones, S., 68

Kennealy, M., 78
Kent, T., 78
Knowledge Economy, GIFT theme, 34, 63, 73
knowledge exchange
  collaborative space, 43
  evolutionary nature of, 59–60
  green learning community, 30–2, 36n4
  SME literature, 58–60
knowledge-transfer activities
  academia and SMEs, 58–60, 62, 86
  business-led maxi projects, 63–73
  business-led mini projects, 73–7
  collaborative research projects, 60–80
  developing curricula for sustainable education, 80–6
  eco-trolleys, 63–5
  input-output model for collaborative research, 62
  integrated stakeholder approach, 59in-situ brokers/ champions, 61
  knowledge brokers, 61
  practitioner-academic partnership, 60–3
  wood fuel quality, 78–80

Lane, E., 68
learning community
  annual learning showcase and study tour, 51–3
  knowledge exchange, 30–2, 36n4
  learning sets, stakeholders, 30, 36n4
  learning spaces, 30
  Learning Showcase Event (LSE), Wales and Ireland, 51–53

McDonald, M., 51, 81, 84
Mann Engineering Ltd., eco-trolleys, 63–65, 80
MBA (Master's of Business Administration)
  greening the MBA, 81–4
  Jones, R., 104–6
  student pitch to small-to-medium enterprise (SME) owners, 76–7
  methane emissions, dairy cow diets, 65–8
Moodle, 46, 75, 98
multilevel engagement, exchange of experiences, 42–3

NESC (National Economic and Social Council), 22n2

Oceanics Surf School & Marine Education Centre
  green audit for, 70–3
  participating SME, 100–2
OECD (Organization for Economic Cooperation and Development)
  Environmental Outlook, 13–14
  Green Growth Strategy, 17
  Working Paper Series, 18
Owens, E., 48, 51, 78, 84, 92, 97

policy gap, energy efficiency and renewable energy resources, 17–20
Problem-Based Learning (PBL) design, 73, 75

Ramblers Worldwide Holidays, 49, 50
region, definition of, 3
Reinl, L., 97

resource based view, 29
Russell, Shaun, 93–5
Russell-O'Connor, J., 70

SDC (Sustainable Development Council), 18, 22*n*2
Small Enterprise Management, Bachelor of Science (BSc) in, 73–5
SMEs (small to medium-sized enterprises)
 changing economic environment, 14–15
 collaboration, 30–2
 enhancement of green innovation in regional, 5–7
 framework for sustainable development, 32–5
 green economy transition in regional, 20
 green innovation and capability, 28–30
 knowledge exchange, 30–2
 MBA student pitch to SME owners, 76–7
 participating SMEs in GIFT team, 100–5
 regional SMEs in green economy, 3–5
 see also knowledge-transfer activities
stakeholder eco-system, cross-border engagement, 34
stakeholders
 co-education ethos, 61–2
 expert/industry, of GIFT team, 99–100
 learning sets, 30, 36*n*4
 SME engagement, 28–30
 see also GIFT team
Storey, S., 65
supermarket retailers, eco-trolleys, 63–5
support hub, GIFT team, 97–9
sustainable development, proposing regional SME framework for, 32–5
 contributory reciprocal cycle, 34–5

tourism
 Bryn Bella Guest House, 68, 102–5
 sustainable walking, 48–51
 training days

sustainable walking tourism, 48–51
wood fuel quality, 78–80
Tuohy, Linda, 100–2

UK (United Kingdom), 2, 8, 49, 50, 64, 68, 71, 78, 80, 83, 99
UNEP (United Nations Environment Programme), green economy, 2
University College Dublin (UCD), 65–8, 92

virtual learning environment (VLE), 32, 46–8, 75

Wales, 7, 15, 45, 83
 eco-certification in hospitality industry, 68–70
 Edge of Wales Walking, 49, 50
 environmental policy, 18–19
 GIFT team, 92–106
 green business landscape in, 21–2
 green technology postgraduate study tour, 84
 Learning Showcase Event (LSE), 51–3
 Well-being of Future Generations Bill, 18
Wales Environment Research Hub, 93
walking tourism
 marketing, 46–8
 training days promoting, 48–51
Wall, J., 46
Walmsley, J., 78, 79, 84
Walsh, M., 63
waste management, 21, 32, 33
 community of practice ethos, 35, 46–7
 future direction, 107
 GIFT theme, 34, 52, 63, 73–5, 94
 Oceanics Surf School & Marine Education Centre, 72
Waterford Institute of Technology, 73, 92, 94–5
Well-being of Future Generations Bill (Wales), 18
wood fuel quality, collaborative engagement, 78–80

Young, E., 48

The manufacturer's authorised representative in the EU is Springer Nature Customer Service Centre GmbH, Europaplatz 3, 69115 Heidelberg, Germany. If you have any concerns regarding our products, please contact ProductSafety@springernature.com

Printed and bound by CPI Group (UK) Ltd, Croydon, CR0 4YY

23/03/2026

02076355-0014